D0609197

What Is a Person?

Contemporary Issues in Biomedicine, Ethics, and Society

What Is a Person?

Edited by

Michael F. Goodman

*Department of Philosophy, Humboldt State
University, Arcata, California*

Humana Press • Clifton, New Jersey

To *Hollie*

Library of Congress Cataloging-in-Publication Data

What Is A Person?
      (Contemporary issues in biomedicine, ethics, and society)
      Bibliography
      Includes index.
      1. Self (Philosophy) 2. Abortion—Moral and ethical aspects. I. Goodman, Michael F. II. Series.
BD450.W4873   1988   126   86-21487
ISBN 0-89603-117-9

Contents

Preface

The idea for an anthology on personhood grew out of two things, viz., the work I did with Martin Benjamin during the Summer of 1982 at Michigan State University on the question, What is a person?, and the amount of time, effort, and expense required for serious research on the topic itself. The former experience taught me the importance of, among other things, attempting to get clear about what we are to mean by 'person,' while the latter experience suggested a possible course of action whereby getting clear might be made more manageable simply by having relatively convenient access to some of the most insightful and stimulating writings on the topic.

The problems of personhood addressed in this book are central to issues in ethics ranging from the treatment or termination of infants with birth defects to the question whether there can be rational suicide. But before questions on such issues as the morality of abortion, genetic engineering, infanticide, and so on, can be settled, the problems of personhood must be clarified and analyzed. Hence *What Is a Person?* has as its primary theme the examination of various proposed conditions of personhood. What qualities/attributes must a being have to be considered a person? Why are those qualities, and not others, significant?

Although this volume was not designed primarily as a textbook, it will be appropriate and of great assistance to both students as well as instructors in any course treating issues in medical and/or practical ethics, where there is concern over any aspect of personhood, humanhood, and decisions to be made bearing on the definitions of these terms. Although a large percentage of the texts currently used in courses in moral philosophy include *some* discussion of the concept of personhood, there has been no book available devoted solely to this topic. The present volume provides more than a mere glimpse at the subtle complexities involved in the problems of personhood.

Considering its scope, the question of who is to count as a person is perhaps the most important issue in ethics today. One aim of this volume is to allow for sustained analysis of the many theses held by writers concerned with the problem. However, despite the fact that a majority of the essays included here focus in some measure on the abortion issue, this is not a book about abortion *per se*. The status of the fetus is the typical "hard case" taken up by just about everyone when the question of personhood arises. Conversely, when the problem of the status of the fetus arises on its own, the question of personhood and humanhood invariably comes up. As will be readily seen, however, who shall count as persons has revelance far beyond the narrow confines of any one or two interesting trouble-spots in what we now call practical ethics.

I owe a great debt to Martin Benjamin for many discussions wherein my dogmatism was exposed. For their understanding, advice and criticism, I wish to thank Richard Anderson, Howard Brody, Bruce Miller, James Roper, Ronald Suter, and Harold Walsh. I would also like to express my deepest gratitude to Don and Karen Bale for their care and encouragement. But the most important acknowledgment I have to make here is to Herbert E. Hendry and Stanley Mortel, my teachers and friends.

Michael F. Goodman

Contributors

Lawrence C. Becker • *Department of Philosophy, Hollins College, Hollins College, Virginia*

Sissela Bok • *Cambridge, Massachusetts*

Baruch A. Brody • *Director, Center for Ethics, Medicine & Public Issues, Baylor College of Medicine, Houston, Texas*

Robert Brungs • *S.J., Institute for Theological Encounter with Science and Technology, St. Louis, Missouri*

Daniel Callahan • *Director, The Hastings Center, Briarcliff Manor, New York*

W. R. Carter • *Department of Philosophy and Religion, North Carolina State University, Raleigh, North Carolina*

Daniel Dennett • *Department of Philosophy, Tufts University, Medford, Massachusetts*

H. Tristram Engelhardt, Jr. • *Center for Ethics, Medicine & Public Issues, Baylor College of Medicine, Houston, Texas*

Harry Frankfurt • *Department of Philosophy, Yale University, New Haven, Connecticut*

Peter A. French • *Department of Philosophy, Trinity University, San Antonio, Texas*

Robert E. Joyce • *Department of Philosophy, Saint John's University, Collegeville, Minnesota*

Edward A. Langerak • *Department of Philosophy, St. Olaf College, Northfield, Minnesota*

Roland Puccetti • *Department of Philosophy, Dalhousie University, Halifax, Nova Scotia, Canada*

Michael Tooley • *Department of Philosophy, University of Western Australia, Nedlands, Western Australia*

Roslyn Weiss • *Department of Religion and Philosophy, Hood College, Frederick, Maryland*

Acknowledgments

Daniel Callahan, "The 'Beginning' of Human Life: Philosophical Considerations," *Abortion: Law, Choice and Morality* (Macmillan, 1970). Reprinted with permission from the author and publisher, and special thanks to Gene Panhorst and Daniel Callahan.

Lawrence C. Becker, "Human Being: The Boundaries of the Concept," *Philosophy and Public Affairs*, **4**, 4 (Summer 1975). Reprinted with permission of Princeton University Press.

Michael Tooley's "In Defense of Abortion and Infanticide" (1984), is reprinted, in its entirety, with the author's permission.

Roslyn Weiss, "The Perils of Personhood," *Ethics*, **89** (October, 1978). Reprinted with permission of the author and the University of Chicago Press.

Harry G. Frankfurt, "Freedom of the Will and the Concept of a Person," *Journal of Philosophy*, **LXVIII**, 1, 15–20, January, 1971. Reprinted with permission from the author and the *Journal of Philosophy*.

Daniel C. Dennett, "Conditions of Personhood," *Identities of Persons*, ed. A. O. Rorty, 1976. Reprinted with permission from the author and the University of California Press, and special thanks to Dan Dixon.

H. Tristam Engelhardt, "Medicine and the Concept of Person," is reprinted with the author's permission.

W. R. Carter, "Once and Future Persons," *American Philosophical Quarterly*, **17**, 1, January, 1980. Reprinted with permission from the author and *American Philosophical Quarterly*, and special thanks to Nicholas Rescher.

Robert E. Joyce, "Personhood and the Conception Event," *The New Scholasticism*, **52**, Winter 1978. Reprinted with permission from the author and *The New Scholasticism*.

Sissela Bok, "Who Shall Count as a Human Being," *Abortion: Pro and Con*, ed. Robert L. Perkins (1974). Reprinted with permission from Schenkman Publishing Co., Inc.

Baruch Brody, "On the Humanity of the Fetus," *Abortion: Pro and Con*, ed. Robert L. Perkins (1974). Reprinted with permission from Schenkman Publishing Co., Inc.

Edward Langerak, "Abortion: Listening to the Middle," *The Hastings Center Report*, 9, October, 1979. Reprinted with permission from the author and the Hastings Center.

Roland Puccetti, "The Life of a Person," *Abortion and the Status of the Fetus*, William B. Bondeson et al., (eds.), 1983, D. Reidel Co., Dordrecht, Holland, 169-182. Reprinted with the author's and D. Reidel's permission.

Robert Brungs, "Human Life vs Human Personhood," *Human Life Review*, **VIII**, 3, Summer, 1982. Reprinted with permission from the *Human Life Review*, New York, NY and the Human Life Foundation, Inc.

Peter A. French, "Kinds and Person," *Philosophy and Phenomenological Research*, **44**, 2, December 1983. Reprinted with permission from the author and *Philosophy and Phenomenological Research*, and special thanks to Ernest Sosa.

Introduction

There has been philosophical discussion for centuries on the nature and scope of human life. Lucretius, for example, contends that human life is inexorably bound to conscious experience; that a person's history is to be viewed from this essential consideration. Kant, on the other hand, while failing to outline detailed ontological criteria for humanness, makes the moral claim that all rational beings should be treated, never as mere means, but always as ends in themselves. We may suppose, however, that Kant *was* attributing rationality to humans, and assume also, I think, that rationality serves for Kant as an indicator of moral agency, taken in its broadest sense. Both consciousness and rationality, *qua* attributes of moral agency, will play a crucial role in much of the discussion throughout the readings that follow.

More recently the term "human" has come under fire from a group that advocates a distinction between 'human' and 'person.' Whereas humans have been, in the past, considered as possessing certain moral rights, specifically the right to life, the deepening concerns of anti-abortionists, to take the clear example, have awakened many analytic philosophers to the need for a more perspicuous account of what is meant by these terms. As we know, ascriptions of humanhood have ranged from the aged to the infant, from the full-term fetus to even the zygote, the question remaining throughout, What must a being be like to be human?

We will see that answers to this question have come in various forms, each writer picking out some one, two, or three characteristics deemed essential for humanness. What is more striking to this writer is the recent fascination with the term "person." Most of us have always used 'person,' and still do, though perhaps with more caution, to refer to some human being we know, or see on the street, or whatever. We say things like, "Yes, I know that person," and "Will the person who owns the green Chevy remove it from the driveway?" And though we seem to be referring to humans in utterances such as these, it would sound exceedingly odd to replace 'person' with 'human being' in like contexts, yielding something such as, "I am the human being who phoned about the carpet." On the other hand, there seem to be perfectly ordinary statements and contexts in which 'person' and 'human being' are apparently interchangeable, e.g., "John is sure a *good...*," where either term fits nicely into the space. It is not at all clear to me that many writers have intended to draw much of a distinction between these terms in their published work, though there is increasing care being taken to accomplish this of late, especially in the writings of such people as Michael Tooley, Peter Singer, and Richard Rorty. And if there is to be a distinction between 'human' and 'person,' we must ask what it is to consist in. The answer will no doubt provide us with different answers to the questions, What is a human being? and What is a person?

Consider the following quite persuasive argument:

'A' is composed of human parts
'A' is a living being.
Hence, 'A' is a living human being.[1]

Surely there is an equivocation here on the term "human." There may even be an element of the fallacy of Composition at the core of this argument. Now, as Becker points out, a conceptus is of the human species and it is also a live being. But this does not justify the inference that the conceptus is a human being. And though it may well be the case that a conceptus *is* a human being, the argument above does not show it. What it does show, I think, is that there just might be a difference between being human and being a human being. For instance, human sperm cells are, strictly by hypothesis, human, but they are not human beings, at least not in the ordinary sense of that term.

One plausible response to the query about what a being must be like to be human is to point out that most of the humans we are at all acquainted with have desires, beliefs, frustrations, fears, and so on. These seem to be, so far as we have yet discovered, uniquely human qualities. But just as it would be questionable to deny humanhood to a being (born to human parents) who did not display one or more of these qualities, so it is equally questionable whether to ascribe humanhood to being (*not* born to human parents) who *does* display one or more of these qualities, i.e., if these qualities do indeed account for at least part of the criteria for being considered human. Again, the natural questions to ask are, What are the criteria for humanhood? and, Are the criteria, or any one criterion, necessary and/or sufficient?

It will appear somewhat far-fetched to many that a fictional character such as Frankenstein's miscreant might seriously be the object of debate over ascriptions of humanness.[2] Humans, for the most part, have rights, obligations and privileges, and are subject to various positive and negative social and legal sanctions, such as taxation, punishment, and so on. Surely, it might be argued, only human beings are taken to be part of the "fellowship of mankind." Anyone can see, just by the utterly primitive nature of his ability to act in accord with the laws of the community, that the "monster" is not a member of *our* community. But consider, for instance, the more realistic case of Charles Manson and his followers. Their acts of slaughter are intolerable—some would say inhuman. Does that make Manson nonhuman? It does not, for now an important further distinction can be drawn, namely, between the biological and the social.[3]

To make this bifurcation have much force, what one needs are clear-cut examples of beings who fit into one category and not in the other. However, before these examples can be enumerated, if there indeed are any, one needs to set out, very carefully, the criteria for belonging to one or the other category. This amounts to defining the terms "biologically human" and "socially human," and indicating how these terms can have different referents.

Let us follow John Noonan momentarily and say that a biological human is a being with a genetic code of a certain kind.[4] The sort of code required, then, for biological humanness will consist in one which is generically identical with that of *Homo sapiens*. Thus, any being not in possession of genetic coding of this nature will be nonhuman. We see that Manson is probably biologically human on this criterion. But, then, so are sperm cells and patches of skin tissue taken from members of our own species. It is clear that this criterion will not suffice for defining 'human being,' but may serve perhaps only for designating what is of the human species.

Thus far there has been no indication that biological humanness is a sufficient condition for what I earlier imputed to *most* humans, i.e., rights, duties, privileges, etc. In fact, individuals imprisoned in the United States for felonious crimes lose many of the rights they previously enjoyed, even though they continue to be humans in what we have called the biological sense. They do not lose all of their rights, however. They retain rights not to be tortured, to be given adequate daily portions of nutritious food, clean air, shelter, warm clothing, facilities for maintaining bodily cleanliness, and so forth. On the other hand, an individual sentenced to capital punishment is in a much different position. Though it is true that all of the aforementioned rights remain in force during the individual's "last days," there is a sense in which those rights are only temporary and seem to be completely absent at a specific moment on the day of execution. Seen in this light, we have a case where a living biological human being has no rights, including the right to life.[5]

If 'biologically human' is reasonably clear and acceptable, at least for our present purposes, our next task is to explicate the meaning of 'socially human.' Let us say that a being is socially human just in case it is a recognized member of the community which includes, but is not necessarily limited to, biological humans. As a member of this public society, the individual must behave in certain "social" ways. A recluse, for example, is certainly biologically human but not, at least most of the time, socially human. Many would argue that the same is true of human fetuses and irreversibly comatose biological humans. One important difference between these cases is that the recluse is, presumably, in such a state by choice, whereas the fetus and the comatose are not.

Notice that on our definition of 'socially human' there is no presumption of biological humanness. The important stipulation is that the social human *behave* in certain ways. Having some specific attributes, such as rationality, free will, an ability for complex communication, and the capacity for self-motivated activity would make it possible for a being to "fit in" and function within the society. So, for the sake of argument, it would be plausible to welcome creatures like Richard Rorty's Antipodeans into our own community.[6] In fact, given Rorty's description of this race of intelligent beings, the *onus probandi* would presumably lie with anyone unwilling to accept them as socially human.

There arises a problem now with the adequacy of the locution of 'socially human.' As mentioned, there is no logical necessity that a being who fits into that category will possess a genetic code defining biological humanness. Hence, there is no necessity that a being who is socially human be, at the same time, biologically human. The term "human" then, in 'socially human,' is somewhat misleading. If perchance beings from different regions of our universe *were* discovered and accepted into the fellowship of our community, even though they were of a quite distinct species with radically differing genetic coding from ourselves and each other, given their capacities for such things as rationality, self-consciousness and complex communication, they would come under the scope of the category "socially human." It seems to me that this would be a dubious application of the term, though not the category. We would hesitate to call these beings human, though we would not, I trust, think twice about calling them *persons*. Oddly enough, this is precisely the epithet Rorty employs in describing his Antipodeans. If they had sensory organs in some sense like our own, we would accompany them to the theatre, admit them into our universities as students and professors, brake our automobiles when they were in crosswalks, and so on. I suggest that we would begin to treat them like we treat each other, i.e., as having a common bond by virtue of living in the same community. Each would be "one of us."

With the introduction of the term "person" here, through the improbable example, we begin to see the outlines of a problem that simply won't go away. How is 'person' to be defined? What are the criteria for personhood? Recall the mention of the dilemma over the status of Frankenstein's creature. At that point, the question was, "What are the criteria for humanhood?" Now, if this question translates into, When is a being a biological human?, the answer is fairly clear if we accept the genetic code criterion. However, in many of the writings of the last twenty years employing the term "human," that term can be replaced with 'person' at all occurrences and lose very little in the translation. And it seems implicitly agreed that 'person' can best be applied to beings who have, as Tooley says, a serious moral right to life.

The problem that won't disappear becomes now the problem of who has a serious moral right to life. And the answer to this question will provide us with sufficient material for answering the question, What are the criteria for personhood? This is so because, if we take the majority of writers on personhood seriously at all, the conditions of personhood are sufficient for ascribing the right to live.

In the papers reprinted in this volume, each author attempts to give some evidence for and against various attributes as relevant for personhood and/or humanhood. Some of the requirements cited are quite unique, e.g., Daniel Dennett's *verbal communication* and Harry Frankfurt's *second-order volitions*. Other qualities suggested as necessary for personhood are more common, e.g., *capacity for rationality*, *awareness of self*, and the *ability for self-motivated activity*. One thread running throughout is the desire and attempt being made to settle the question about who will be/should be accepted into the fellowship of the human community, with all the rights and privileges applicable to that membership. Dogs, for instance, have not been accepted into our society as persons, and are designated nonhuman nonpersons. The species *Canis familiaris* has, however, had certain rights conferred upon it, or, better, we have in some sense obligated ourselves to treat members of this species in certain ways and to refrain from treating them in other ways. The standard for such treatment seems to be covered by the meaning of the term "cruelty." Dogs ought not be tortured, deprived of food and water, or be exposed to excessive doses of heat or cold. The list could continue. Nonhuman nonpersons are accorded a due measure of concern by us for their well-being, but this does not include the right to live, or the claim that heroic measures ought be taken for their benefit. This concern is not moral concern except perhaps in so far as a principle of general utility might require its being so taken in some cases.

Human persons, on the other hand, *do* have the right to live, at least *prima facie*, the important question being, What is the extension of 'human person?' We are at a distinct advantage here, as there are some clear-cut cases of human persons, e.g., Mickey Mantle, Margaret Thatcher, Alexander Solzhenitsyn, James Earl Jones, me, you, and the common "man on the street." Two hard cases are fetuses and those in a state of irreversible coma—human, no doubt, but persons? We will encounter a number of proposals for placing these cases in what might be called the human nonperson category. This will amount to something analogous to nonhuman nonperson status in that life rights will be denied though a great deal of [moral] concern will be expected of us for these individuals.

I think the human nonperson class will be thought by many to be strictly empty, something impossible—much like the set of synthetic *a priori* judgments accepted by Kant and a handful of others but merely rejected by the majority. For many people, the terms "human" and and "person" are interchangeable. But something of a revolution (in the Kuhnian sense) in the thinking about personhood has taken shape over the past decade disallowing the synonomy of these terms.[8] One line of argument here might be that though the fetus is human, with a human genetic code, since it does not possess any of the conditions for being a person, here and now, such as self-consciousness and rationality, it cannot be considered a person with a serious right to life. Hence, a seemingly clear-cut case of human nonpersonhood. An analogous argument is many times made these days regarding those individuals in irreversible coma, the paradigm instance being that of Karen Quinlan.[9]

But this is not the only "shocking" category derivable from the combinations of 'human' and 'person.' Let us consider the dolphin. Studies are proving that this cetaceous mammal is quite intelligent, and it is not inconceivable that the mental capacities of one of these adult creatures are more mature than an adolescent human person. If the dolphin were shown to be rational, self-conscious, able to adopt and reciprocate an attitude of moral concern for another being, in short, if the dolphin were shown to possess all of the relevant conditions of personhood, i.e., all of the qualities we consider relevant for ascribing personhood to ourselves, we would be charged with speciesism and/or self-contradiction if we intentionally failed to recognize them as persons. And so our fourth category, that of nonhuman persons, not only becomes initially plausible, but something over which our thoughts might linger for a richer understanding of how we view the "lower" forms of animal life.

It is not difficult to understand how a person immersed in Western culture could find the concepts of a human nonperson and nonhuman person unnatural and confusing. Since we are weaned on the ideas that pain is to be avoided for the most part and that human life is precious, our thoughts rarely stray, outside philosophy, from the seeming interchangeability of 'human' and 'person.' Humans are persons and persons are humans; that's the way it is and that's the way it will be. Unless, of course, we change things. And that is what much of this book explores.

II

The first and second papers in this collection are intended to serve primarily as introductory materials for clarification of certain conceptual difficulties to be found in a number of papers that follow.[10]

In Daniel Callahan we find an awareness that seems to be lacking in too many writers on the aspects of personhood. It is that human beings are more than biological entities, endowed with certain moral value in virtue of this uniqueness. Human adults are more than merely capable of self-motivated action and complex communication, and possessed of an opposable thumb. Humans are physical, psychological, and cultural creatures, and it is the unity of these areas of our existence that ought to be shaping our thoughts and decisions when questions in medicine, law, and morality are confronted.

When does human life begin?, is Callahan's initial question. But in some sense, he says, we clearly know. We really couldn't even ask such a question, using the operative terms "human" and "life," without at least knowing how we normally use these terms, and further, without knowing what we generally mean by them. However, the intimation that we already know what we mean by 'human' seems only as much as to say that we have some clear examples of human beings, viz., "fully developed men and women." "To be human," Callahan writes, "means to possess those characteristics we associate with the word 'man.'"[11]

As is pointed out, there is no dearth of historical material on the essence of humankind. From Aristotle to Aquinas and Descartes to Mead we have lists of the essential characteristics of *H. sapiens*. From more recent anthropological and zoological studies we have even further considered opinion on the "nature of man." What

Callahan asks us to do, rather than pick out this or that feature of being human as salient, is to consider the significance of the biological, psychological, and cultural interrelationships ever-present in every human being. To define 'human being' strictly from a psychological viewpoint, say, will yield a severely impoverished theory of "man." This will be so no matter what single area bears the weight of the definition.[12] If it is a definition of 'human being' we seek that will be useful for reflecting upon moral questions relative not only to fetus, neonate, adolescent and the aged, but to meaningfulness and quality of life dilemmas, only a rich, comprehensive definition resulting from consideration of the innumerable cross-disciplinary worlds in which we live and find ourselves will suffice.

In stark contrast to this wide-scoped view of the nature of humanhood, Lawrence Becker is concerned to delimit the *biological* boundaries of human life. On the one hand there is the being/becoming boundary, where he points out that a human fetus stands in roughly the same relation to a human being as a caterpillar stands to a butterfly. The fetus is not the human being it will become, he says. Rather, it is a human becoming. A second consideration Becker points to is the being/has been boundary, where "... a human organism is dead when, for whatever reason, the system of those reciprocally dependent processes which assimilate oxygen, metabolize food, eliminate waste, and keep the organism in relative homeostatis are arrested in a way which the organism itself cannot reverse."[13] At both ends of the life spectrum, then, the determination as to when the status "human being" has been achieved and lost can be made solely on biological grounds. In what follows I shall only concern myself with Becker's becoming/being boundary.

Humans don't just pop into being. One becomes human through a process. In fact, the entry into the class *is* the process. This, I take it, is the central thesis of Becker's paper. But now, what is this process, i.e., in what does it consist? Briefly, what we have here is a metamorphosis from human sperm and ova to fetus in at least the sixth month of gestation, at which point (approximately) the metamorphic phase of generative development can be said to be at and end (conservatively). The metamorphic process consists in establishing the basic structures of the organism and is complete when the organism's basic gross anatomical form is in place *and* when the organism possesses a complete set of histolically differentiated organs. Here, Becker is quick to point out that these criteria are much more easily confirmed conceptually than empirically, though I might add that, with continued striking advances in medical technology, our empirical knowledge might well overtake our conceptual certainty in this, and related, matters.

At the end of the metamorphic phase, then, a human being exists, whereas prior to its completion what we have is a creature who is a member of the human species but not, yet, a human being. This is not, however, a moral divide, says Becker. For that we need some sort of definition of 'human being' that incorporates the concept of personhood in a way that yields "rights to life." Merely being a member of the human species does not produce that result. Nor does entry into the class of human beings. With Becker we might ask, What *morally* significant distinction is captured by the completion of the metamorphic process of generative development? The rule against homicide, for example, seems to require either a victim-based or an agent-based approach, and the becoming/being boundary does not, in itself, seem to embody the critical link essential for either of these approaches.

In a number of essays in this volume, the notion of potentiality will be discussed, e.g., Tooley, Joyce, and Langerak. Extreme anti-abortionist conservatives, of course, argue that the fetus is a person and, pace Becker, not just a human becoming/being from the moment of conception. They appeal to some sort of general principle of potentiality, such as the following, where 'c' refers to either a sufficient condition or a set of sufficient conditions for personhood:

> All and only those creatures who either actually or potentially possess c (that is, who either have c now or would come to have c in the natural course of events) are moral persons now, fully protected by the rule against homicide.[14]

Neither Callahan nor Becker appear to have much sympathy for such a rigid principle. In fact, Becker, like Tooley, would seem to be pointing to the logic of the situation by intimating that, *ex hypothesi*, a potential person is not a person, otherwise there is no force in differentiating the two. For all that, the less than crisp line between becoming and being a human being might make us pause and return to Callahan's essay for a moment.

He writes, "To be 'human' is not just to display, here and now, the full range of human characteristics."[15] We might say the same, of course, of persons. Human persons are rational sometimes, irrational other times; reminiscent *and* forgetful; sublime *and* base; conscious, subconscious, unconscious, *and* self-conscious. A sleeping adult human, for instance, is in a state of temporary unconsciousness. But neither does this being lose membership in the class of human beings nor cease to be protected by the rule against homicide. A sleeping adult is still human and still a person. Human life, says Callahan, is bound up with potentialities and capacities. To define "man," the entire career must be accounted for and this includes, it seems to me, what humans have been, what we are, and what we may become.

The interest in potentialities and capacities is taken up quite forcefully in Michael Tooley's essay. His major thesis is that fetuses and newborn infants (neonates) do not have a right to life. This is so even though they have the potential to have such rights, i.e., they are potential persons. The main thrust of Tooley's paper is that only those beings that can have desires and interests can have rights, including the right to life. His argument is that:

> What is in a thing's interest is a function of its present and future desires, both those it will actually have and those it could have. In the case of an entity that is not presently capable of any desires, its interests must be based entirely upon the satisfaction of future desires. Then, since the satisfaction of future desires presupposes the continued existence of the entity in question, anything which has an interest which is based upon the satisfaction of future desires must also have an interest in its own continued existence. Therefore something which is not presently capable of having desires at all—like a zygote—cannot have any interests at all unless it has an interest in its own continued existence.[16]

It seems to me that this line of argument, coupled with his polemic that a zygote cannot have an interest in its own continued existence, yields the compelling conclusion that a zygote has no right to life. But it speaks not only to the case of the zygote. It indirectly refers to the fetus, the neonate, and all those who are not presently capable of any desires and who have no interest in their own continued existence. This, I suggest, leaves little doubt as to the status of the potentiality principle. Merely having the potentiality for having desires and interests is clearly not sufficient for possessing a right to live.[17]

Now if potential persons are not persons yet, who are persons? What attribute(s) must a being possess in order to be a person? Oddly enough, Tooley shys away from the term "person" in this essay. In his previous papers, and in his book, he makes much of the distinction between humans and persons, where he states that 'x is a person' means 'x has a serious moral right to live.' Having a right to live is bound up with possessing a special sort of self-consciousness, i.e., the being must possess the concept of a self as a continuing subject of experiences and other mental states; it must itself be such an entity, and it must believe that it is itself such an entity.[18] The importance of the concept of a continuing self (or mental substance) for Tooley is further established by its centrality in a number of premises of the argument found in the present essay on the necessary conditions an entity must satisfy for possession of a right to life.

The force of Tooley's arguments for abortion as well as infanticide can now be seen. Neither fetuses nor neonates have this special self-consciousness. Hence, they are not persons. This, of course, does not bar them from membership in the species *H. sapiens*. What Tooley's argument does show is that there is a distinction between humans and persons, and that it is permissible, contrary to wide public opinion, to take the lives of some innocent human beings.

Roslyn Weiss, however, is not happy with the recent trend in the right-to-life debate that advocates using 'person' rather than 'human' to denote that which has such a right. One unfortunate problem with this usage, she notes, is that it leads, in the cases of Tooley and Warren,[19] to taking infants as having no more life-rights than fetuses, since neither satisfy the proposed conditions of personhood. But more than this, even if the advocates of infanticide on these grounds could somehow defend the "rights" of sleeping adult humans and those who are temporarily comatose, Weiss thinks there is no way to "save" the permanently comatose and the severely retarded. Whatever the advantages, then, of introducing the concept of personhood into the debate, there are distinct disadvantages.

If these considerations are not enough to hold back the tide carrying the new terminology into practice, Weiss' next objection may do some sandbagging. Taking "personhood" as the basis for having moral rights, we are left with a rights-based morality. But this is insufficient to account for our conception of the importance of duties, especially in those cases where we affirm the existence of duties that do not correspond to anyone having rights. As an example, one might consider, contra Tooley, that kittens have no rights whatever, but that we have the duty not to cause them unnecessary pain. A morality based solely on rights would preclude any duty to "lower" animal life, dead people, country, etc. And even if the fetus is not a person, she says, its innocence might have a significant role to play in the abortion controversy, since, "It is likely that our duties toward noninnocents are not as great as those toward innocents."[20]

Harry Frankfurt's essay is one of the most influential to ever appear in print on the notion of personhood. Part of its importance lies in the fact that Frankfurt points out a truly remarkable condition for personhood. But not only is this condition remarkable, it is plausible that it comprises a necessary and sufficient condition for being a person. Another point in its favor is that the essay does not stipulate that only humans can be persons. Whereas Frankfurt does indicate that the condition he has in mind does seem to be a peculiarly human characteristic, he also notes that what personhood requires is not a set of attributes that is necessarily species-specific. So, I take it that if we come to discover that the dolphin, say, is a bearer of Frankfurt's special condition, we might well want to ascribe personhood in this creature.

So, what is this condition? Simply, it is what Frankfurt calls *second-order volitions*. Consider a drug addict who desires an injection of heroin. The bare desire for the drug (which may or may not involve other desires, e.g., the feel of the tourniquet on the arm, the feel of the injecting the needle, etc.), and the accompanying state of euphoria, is what we might call a first-order desire, and may be taken in the form "A desires to X." A desire of the second order, however, is much more complicated, and this complexity seems to preclude many nonhuman animals from possessing these desires. In a second-order desire, "to X" directly refers to a desire of the first order. In our example, the addict has a second-order desire when he/she not only desires the drug, but, for whatever reason, desires not to desire the drug. The form it might take is that "A desires (or does not desire) to desire to X." It seems reasonable to admit, at this point, that no matter how we may view the value of our pet dogs, cats, goats, and horses, they just don't possess second-order volitions. In this case, if Frankfurt's special condition is accepted, along with Tooley's suggestion that persons are those beings with rights to life, we might look to Weiss' argument that even though rights are not assignable, duties may be.

The final point I should like to make about Frankfurt's second-order volitions is that it links up with another seeming essential condition for personhood, i.e., freedom of the will. It appears to Frankfurt that true freedom of the will is had by an agent whose second-order desires have become *effective*, that is, where the desires of the second order move the agent all the way to action. Where someone merely has a second-order desire, but where that desire is not sufficient to initiate action, we cannot say that the individual has true freedom of the will. Conversely, where someone has second-order desires and wants those desires to be her/his will, Frankfurt terms them "second-order volitions," which indicate motivation to act via freedom of the will. For Frankfurt, a person is a being with second-order *volitions*.

Daniel Dennett speaks directly to the question whether all and only humans are persons. He says that infant human beings and humans declared insane are denied many crucial elements of personhood. On the other hand, he writes, though humans are the only persons we recognize right now, and though the terms "humanity" and "persons" seem to be locally coextensive, we can easily contemplate beings from other planets, who are biologically nonhuman, as persons. This might be reasonable, I take it, if these creatures displayed the sorts of attributes we count as significant, even essential, for conferring personhood upon ourselves.

Dennett lists six conditions for personhood, all of which he takes to be plausible candidates for necessary conditions on some interpretation or other. They are: *rationality, consciousness, the capacity to take a personhood attitude (stance) toward the being in question, the ability to reciprocate such an attitude, the ability to engage in verbal communication*, and *self-consciousness*. He is concerned to argue for three basic theses in this essay. First, how, on his interpretation, the conditions of personhood are dependent on each other; second, in what sense they are necessary conditions; and third, the difficulty of showing that the conditions are, jointly, sufficient conditions for moral personhood.

Dennett makes an interesting distinction. He says there are at least two interconnected notions of personhood, viz., the moral and the metaphysical. The latter is roughly the idea "...of an intelligent, conscious, feeling agent...," while the former is the notion "...of an agent who is accountable, who has both rights and responsibilities."[21] The question arises of course, for Dennett, whether these two concepts of the person coincide. That is, what is the connection between being a person in the metaphysical sense and being one in the moral sense? Is being one a necessary and/or sufficient condition for being the other?

Of the six conditions listed, three are unique. Whereas most writers on personhood acknowledge the importance of some sort of ability to engage in complex communication, Dennett is alone in advocating *verbal* communication. It is not clear to me that he would be entirely happy with interpreting this condition as requiring vocalization. The adult human, for example, whose vocal chords have been severed, but who retains all other necessary conditions for personhood, must surely be viewed as a person, at least in the moral sense. But if vocal communication really is intended here, then perhaps this human ceases to be a person in the metaphysical sense. In that case, the moral and metaphysical senses of personhood really are more distinct than we might otherwise believe.

The other two uncommon conditions in Dennett's set are the adopting and reciprocating a certain attitude toward an entity. I had previously construed this attitude as one involving the idea of personhood itself, whereby a person is a person in virtue of other persons treating her/him as a person. This seems to me now to be not only begging the entire question of the conditions for personhood, but just plainly contrary to fact. Think of your Aunt Penny who sets a place for her pet poodle at the dinner table. Think of the countless persons in history who were treated like dogs. Think of the Jewish population during WW II and the practice of apartheid in South Africa today. There is a certain dignity that goes along with being a person that treating or not treating in certain ways cannot touch.

For all that, however, I do think Dennett hits the mark squarely in positing the ability to reciprocate such treatment (though perhaps we might entertain the idea that the treatment in question be of a morally relevant nature), as a necessary, if not sufficient condition. I can imagine some *Star Trek* creature too hideous to look at. If this being could adopt a moral attitude to fellow creatures, I think I would be willingly constrained to count it, and treat it, as a person.

In the end, Dennett embraces a skeptical position on ever finding *the* necessary and sufficient conditions of personhood, even though, as he admits, the concept of a person may be "...in some sense an ineliminable part of our conceptual scheme."[22] Why? Because the notion is inescapably normative, where any decision to include or exclude this or that sort of being into the community of persons is at least partially arbitrary. It is, finally, just impossible to discover the objectively satisfiable sufficient conditions for taking any being to be conscious or rational.

Tristram Engelhardt's essay takes up a number of themes we have previously encountered. On the one hand, he echoes Callahan's injunction that we take human beings not merely as biological entities when he says that biological life can be distinguished from personal life. On the other hand, he seems to agree with Tooley that self-consciousness might be necessary for a moral order (though Tooley doesn't put it in just this way). On a third hand, Engelhardt distinguishes two senses of the concept person, albeit he says there are probably many more.

This latter point seems to me the essential thrust of the essay. The first sense is used in reference to entities who are self-conscious and rational. It is termed, by Engelhardt, "strict" personhood, and applies to beings who are identified as moral agents, who are "...individual, living bearers of rights and duties."[23] The second sense, "social" personhood, is applied to instances of human biological life who are treated as though they are persons when in fact they do not meet the conditions for strict personhood. The example used in this essay is that of the mother–child relationship, where the infant, who is neither rational nor self-conscious, and hence not a person strictly, is treated as a person. The infant is socialized, and treated *as if* it had wants and desires, where its cries are interpreted in ways actually appropriate only for strict persons. The social sense of person applies equally well to those who are senile, severely retarded, and the mentally infirm, who cannot be taken to be persons in the strict, moral sense.

Notwithstanding my comments on Dennett's third condition, I concur with Engelhardt's reasons for positing the social sense of personhood. Whereas Tooley's logic for not counting fetuses and infants as persons seems impeccable, there is still an extremely uncomfortable feeling that attends accepting, even theoretically, the permissibility of infanticide. Our realization of the importance of the social role infants play seems essential for the retention of the moral fabric of the community in which we find ourselves.

But more than this, as Engelhardt says, we find infants already playing the role of persons. It is true that young children do live in and through their families. But it is also true that they do, at times, see themselves as themselves; not of course with the advanced self-consciousness of adults, but with the primitive mind of one who is learning to do so. No doubt young children, infants, and neonates are not strict persons. But if we find the line between nonperson and person not as sharp as we would like, perhaps Engelhardt's social sense of person has a place not only in our conceptual scheme, but in our practical lives as strict persons as well.

W. R. Carter's essay seems to me to take off with a notion that was only implicit in Engelhardt. An infant is a social person, according to the latter, but not a strict person. Essentially, an infant is sort of a person and sort of a nonperson, caught in the gray area of our undecidibility. They are not, on Tooley's account, protected by a rule against homicide, but, on Engelhardt's view, they deserve to be taken into our moral/social community. The focus of Carter's essay has to do with degrees of personhood.

The basic thesis of this paper is that since personhood is closely connected with the notion of having certain *capacities* (perhaps rationality, self-consciousness, and second-order volitions), and certain *rights* (perhaps the right to life), and since it seems reasonable to say that these capacities do not all come to be had at once, and neither does any one capacity exist from its inception in its final form, we can conclude that (1) there are no such things as essential persons, (2) the line between being and not being a person is not arbitrary, and (3) there are such things as marginal persons. Each one of these claims can be seen to be important for our understanding of the concept of personhood. Where nothing can be said to be an essential person, we have to rethink the notion of "I." If there was a time when I did not possess the conditions for personhood, and if there will likely be a future similar time, what am I? Further, if the right to live is intimately bound with being a person, as Tooley says, and if I am not by nature a person, whatever right to live I now possess is not an inalienable right. In other words, I can still be, and be me, and have no right to live.

The second point, about the nonarbitrariness of being or not being a person, seems to hinge on being able to discover when an entity possesses the relevant conditions for strict personhood to a sufficient degree. This in itself presents problems, since it is not clear whether personhood is something to be actually discovered or merely decided upon by a consensus of the members of any one given community. As Carter clearly sees, if being a person is to be based on arbitrary decisions, lines may be drawn differently in New York and Moscow.

The thesis that there may be such things as marginal persons, though plausible, solves some problems and creates others. One question that won't arise is whether marginal persons are persons. They are persons, though not strict persons. As persons, however, whereas they will carry at least *some* moral weight, it will not be on a par with the moral status of strict persons. Hence, in any question of who has a greater right to scarce medical resources, or exotic treatment for prolonging life, the claim of the strict person overrides that of the marginal person. Moreover, since marginal persons are persons, and since the obligation of persons toward persons overrides a person's obligations toward nonpersons, the decision to save the life of a marginal person rather than even a beloved pet dog will not be difficult to take.

The one most difficult problem arising from positing degrees of personhood seems to me to be that of determining just who are marginal, and to what degree. Clinical situations, real and imagined, that would benefit from such a determination are innumerable. What degree of brain activity, for example, is to be considered sufficient for ascribing more than marginal personhood? Is Karen Quinlan, in a state of irreversible coma but with some brain activity, a marginal person or a nonperson? It is not clear that there is much we can do to *discover* answers to these questions, but like Carter, I shy away from the arbitrariness of merely *deciding*.

If Tooley can be seen to have taken on a radical position regarding personhood, and if Carter and Engelhardt are seen as no less extreme in their polemics, what Robert Joyce has to offer in his essay matches each of these writer's immoderate stance on who shall count as a person. The essential difference is that Joyce argues for the personhood of human zygotes from the moment of conception, a position known as extreme conservatism.

Most people who agree with Joyce's central thesis make use of some kind of principle of potentiality. This is not necessarily so, however, for those who advocate some version of ensoulment criterion, where an entity becomes a person at the moment of possession of a soul.[24] But Joyce's thesis does not involve the notion of ensoulment. Rather he makes use of an idea of potentiality that embodies actuality. In short, the capacity (potential) to engage in a certain practice (action) that one is not at present engaged in is an actual potential, or, a functional capacity. Since I have the ability to ride a bicycle, but am not presently doing so, we say I actually have the capacity to ride. This was not so when I was, say, two years old. At that time I had the capacity to learn to ride rather than the capacity to ride.

But now consider various previously proposed conditions of personhood. Certainly a zygote does not possess the actual capacity for reasoning or being self-conscious. At this point Joyce makes a crucial distinction regarding capacities, citing Callahan as well as the *Roe vs Wade* decision as taking a "developmentalist" position on the beginning of the human person. This view is that a person is an entity that has a developed capacity for reasoning, willing, desiring, and so forth. Joyce remarks that a person is not someone who has already developed the abilities "required" for personhood, but is one who has the *natural* capacities, whether they are ever developed or not.

I suggest that this attitude is not without some support from other writers in this volume who would not accept the personhoood of a conceptus. For example, Engelhardt does say that it is perfectly plausible that one might take a fetus to be a social person at viability. And Dennett's third condition, adopting a personal attitude with respect to the being in question, seems to allow for us treating a being as a person whether it can reason at the present moment or not. Further, Joyce does not violate the logic of adopting a version of the potentiality principle by referring to the conceptus as a potential person and also as a person. As mentioned, a potential person is, *ex hypothesi*, a nonperson. Perhaps with that in mind, Joyce refers to these beings as prenatal persons and prebirth children, all the while implicitly arguing for the actual natural potentiality of the fetus to achieve the functional capacities to display the "conditions" of personhood.

Sissela Bok's essay gives a nice brief exposition of the various opinions on when a fetus becomes human. It seems to me that the notion of humanity, so used by Bok, has given way to the current notion of person, i.e., a being with a right to life in the moral sense, or, in law, a being protected by a rule against homicide.

One interesting aspect of Bok's paper, for our purposes, is the application she makes of the reasons for protecting life to the case of prenatal life. Her reasons for protecting life are:

(1) Killing can be seen as the greatest of dangers for the victim.
(2) Killing brutalizes and criminalizes the killer.
(3) Killing can cause grief and a sense of loss for the victim's family and others.
(4) Society has a stake in the protection of life.

In applying these principles to prenatal life, we contrast the zygote, embryo, and fetus with the adult human being, a clear person. Where each of the stated reasons for protecting the lives of persons does have instances of occurrance, the case of an abortion, desired by both parents, lacks certain crucial factors required for protecting the fetus' life. Some of these factors have to do with the differences between human fetuses and human adults. As Bok points out, a zygote cannot feel the anguish bound to (impending) death; it cannot fear its own destruction; it has not yet begun to experience its own life and it could not be conscious of any interruption of its life. We might say that there is no "for its own sake" here, no "for the victim" as there is in the case where a fully conscious, knowing human adult faces death.

This is not, of course, to say that there may not be other reasons for protecting the lives of fetuses. We can easily see that (2), (3), and (4) above might carry some weight in certain situations, e.g., where the practice of abortion, at whatever stage of fetal development, becomes merely a method of birth control. It must be agreed,

I trust, that Bok's point is well taken that even if the fetus is human, or a person, that factor alone cannot establish a moral imperative to protect its life in every situation. Other people, and their desires, wills, and reasons, are involved.

Baruch Brody's essay ought to be read, perhaps, by every freshling in every course on contemporary moral issues. The problem is the humanity of the fetus and when it can be said to be a member of our moral community. As with Bok's essay, I think we could replace 'human' with 'person' and lose little by doing so. In fact, some insight might be gained since, as I mentioned, current writers on the topic need not take 'human' as a term denoting life-rights whereas everyone seems to take 'person' as having normative connotations.

Brody examines the various stages of fetal development to determine when, if at all, the fetus becomes human. He lists no less than twelve arguments, each advocating a different stage for fetal humanity, from conception, when a human genetic code exists, to some time *after* birth, when the possibility of interaction with other humans exists. Finally, with all but one argument rejected, Brody fixes on "the essence of humanity," i.e., that attribute/property which no human being can lose without going out of existence—brain activity. Used as a criterion for death, the lack of brain activity is an indication that the being in question is no longer a living human being. Used as the criterion for humanity, Brody concludes that it is the essential feature of all living humans, including, obviously, the fetus.

The position Brody adopts here has some clear relevancies for Carter's marginal person and Engelhardt's social person. Or perhaps it's the other way around. According to Brody, it may be the case that a six-week fetus is fully human, endowed with all the rights and privileges accompanying such status, including a claim to life equal to the claim of its mother. And *that* is a person in the strict, nonmarginal sense.

Edward Langerak's essay has a number of interesting points to consider. In the first place, he advocates a version of the potentiality principle that makes abortion at least morally problematic; not as morally prohibitive as Joyce would have it, of course, because Langerak does not argue for the personhood of the fetus. Rather, it is something about the fetus, as a living human being, though a *potential person*, that makes it morally problematic to take its life.

This something is the potential a fetus has for attaining actual personhood, where an *actual person* is an entity possessing "...a sufficient condition (whatever that may be) for personhood and thereby has as strong a claim to life as normal adult human beings."[25]

Second, Langerak makes a number of distinctions between different kinds of persons and nonpersons. The two types of person mentioned are the actual person and those who have a *capacity for personhood*. For instance, sleeping adult humans are persons in virtue of having reached a stage of development where they *could* display the conditions of personhood, even though they are not now doing so. As examples of nonpersons, Langerak lists the human fetus as a potential person and a human sperm or egg as a *possible person*, i.e., a being that might, "under certain causally possible conditions," achieve actual personhood.[26] This distinction between potential and possible persons is indeed an important one, for Langerak argues that whereas the former have a claim to life, the latter do not.

This is an interesting thesis to me, i.e., that a nonperson might have a claim to life, because throughout this introduction I have been making the point that only persons can have such a claim. Langerak bases his argument on the notion that there is some intimate relationship between future claims of an entity and present claims. This point, he says, rests on the fact that we generally view humans in terms of their temporality, in terms of projecting ourselves into the future from the places we are at now. And the fact that we put so much stock in these temporal projections might well lead us to respect humans for their potential. This seems to me a plausible thesis for arguing that a society concerned with the future and the potential of its subjects ought to confer at least a *prima facie* claim to life on potential persons, even with the stipulation that they are actually nonpersons.

Roland Puccetti opens his essay with a fascinating tale. A genie appears to you and says that, if you want, he will expand your brain in such a way that you will have an IQ of 400. You will become an Einstein, a Picasso, or a Pasteur. There is, however, one drawback. At the moment of expansion, your conscious experience will cease forever. Your brain will be programmed to achieve great things but, as Puccetti puts it, you will be an "automaton" and nothing more. The question: Would you accept the genie's offer? Well, perhaps as an altruistic act that might result in discovering a long-needed cure for some dread disease, one just may consider it. But, we suspect that the loss of consciousness would be an overriding factor in many people's rejection of the genie's proposal.

This, I think, is Puccetti's central point, i.e., that the "I" of a person is thought to disappear with consciousness. And if I disappear, if I go out of existence, surely that person I was also ceases to exist. It seems absurd, then, the argument continues, to extend personhood to beings who have lost the capacity for consciousness. And no less absurd to confer personhood on beings who have not yet attained conscious experience. In just this way, Puccetti seeks to exclude early fetuses (because of their lack of neurological hardware) and those who are irreversibly comatose (because of their loss of *personal* life) from the set of persons.

Puccetti goes on to argue that many people do confuse non-persons with persons and, like Langerak, makes important distinctions between *possible, potential, beginning, actual,* and *former* persons. As we've dealt at length with the other categories, the one that might interest us here is "beginning persons." 'Person' being coextensive with 'moral agent,' Puccetti places human children in this class. He does so since children are moral *objects*, like other forms of "higher" animal life, but are only potential moral *subjects*. A child, then, is sort of a marginal person, at least in the sense that it has come to be considered as part of the moral community. This is not sufficient, however, for ascribing actual personhood. In the end, Puccetti admits to sharing Dennett's skepticism about ever being able to set out an exhaustive list of necessary and sufficient conditions for personhood. For all that, it seems he has given a reasoned argument that consciousness will be on that list somewhere.

Robert Brungs' essay seems to me to display a certain anger and sadness that many writers on personhood share. The majority opinion in *Roe vs Wade* distinguished between human life and human personhood. As a result, "In the name of reproductive *freedom* we have embarked on an essentially *totalitarian* estimate of the human being,"[27] with the state/society now deciding who is to be considered a person, protected by the law, and who is not. Brungs' deep sorrow and indignation stem at least in part from the fact that the unborn child can now be treated arbitrarily, since it is not considered a legal person with any rights whatever.

Because, in large measure, of striking advances in science, technology, and industry, the world has changed radically. In the past, he says, the changes wrought were for our betterment, and without doubt many efforts to make human life better have been entirely successful. However, since the coming of age of the biological sciences we have embarked on an adventure that will ultimately makeover the human body and psyche in such fashion that the day will arrive when humans themselves will be the most important of artifacts. With the enormous advances in medicine, for example, that have already taken place, we can look forward to even greater achievements, not the least of which will be the possibility of "enhancing" our genetic inheritance through eugenics. And at that stage of the game, *humans* will begin to change radically.

We might worry here about two things. First, we would hope that the new sciences and technologies will always have in mind the good of the species, not to mention the enviroment in the broadest sense. But what seems to be presupposed here is that we know what is good for ourselves. For instance, we can imagine the proponent of eugenic research arguing that in the end we will have "better" human beings. We would naturally ask, Better than what, in what respects, and to what degree? The fact is, we don't pretend to know what a "good" human being is, and especially not in the sense in which the art/science of eugenics might be useful.

Second, where science and technology advance to a point of "deep" research on human beings (if this is not already happening), there is the fear that humans, and hence some persons, will become *objects*, used as mere means instead of ends in themselves. This will involve another radical shift, viz., in the thinking of these research scientists and technologists, who are, incidently, human persons themselves. The question, of course, is, What happens to the concept of the person, and the person her/himself, when humans become objects in this way?

Peter French asks the following question: Is 'person' a natural kind term? The answer is a complicated "no." Briefly, natural kinds are best seen as instances of set membership based on sameness of structure and function relative to sets of natural laws.

"Water," on this view, is a natural kind. So is "mallard" and "human being." However, our commonsense uses of 'person,' based in large part on a coherent set of empirical observations and generalizations, does not lead us to affirm it as a natural kind term.

French's essay is a departure from most of the others included in this volume. He is not concerned with the decision of *Roe vs Wade,* or the status of infants and young children. He is at pains, however, to show that 'person' is not necessarily coextensive with 'human being.' If some "man" were to land on earth in a spaceship from a destructing planet far away, and if this being looked like us and acted like us, especially with respect to intentionality in the widest sense, then even though his biological structure was completely different from our own, we would, naturally, call this being a person. We would include him in our community, moral and otherwise.

But French goes beyond the conception that even various sorts of nonhuman animals might be considered persons. He makes a plausible case, I think, for taking 'persons' as applicable to very sophisticated computers/machines, e.g., Hal in *2001: A Space Odyssey*, and business corporations. The crucial reasoning behind these ideas is that deciding that some entity is a person is, at very least, to judge that it makes sense to describe some aspects of its behavior in terms of intentionality on the part of the entity itself. Genuine intentional systems, then, whoever/whatever they turn out to be, are persons.

Epilog

As will be seen, the shared concerns of the papers that follow run deep. It is at once paradoxical and proper that it is persons who ask the question, Who are persons? It is fitting that persons concern themselves with not only themselves but with the worlds in which they are intimately a part. It is not enough anymore to merely count ourselves as essentially human, endowed with intrinsic moral worth. There is a new term on the lips of philosophers, theologians, politicians and scientists that is beginning to compete with 'human being' for a privileged place in our thinking about ourselves and others. That there is no general agreement on either the sense(s) in which 'person' should be taken or its extension ought not trouble us at present. The conversation has been taken up, and it is hoped that the reader will contribute to its continuance, as do the essays in this volume.

Notes

[1]Lawrence C. Becker, "Human Being: The Boundries of the Concept," *Philosophy & Public Affairs*, **4,** 4 (Summer 1975), Princeton University Press.

[2]Mary Shelley's *Frankenstein* has much force and depth as a work of fiction. I think one can gain certain insights into what we mean by 'human' and 'person' by reading it with this in mind.

[3]Daniel Callahan, *Abortion: Law, Choice, and Morality* (Macmillan, 1970), divides the relevant criteria for the determination of human life into the generic, the developmental, and the social. Albert S. Moraczewski, "Human Personhood: A Study of Personalized Biology," *Abortion and the Status of the Fetus,* edited by Bondeson, et al. (Reidel, 1983), criticizes the development and the social. I dispense with developmental merely for convenience, preferring to include it in the social as an essential element in history for thinking about persons and personhood.

[4]John T. Noonan, Jr., "An Almost Absolute Value in History," *The Morality of Abortion,* John Noonan, ed. (Harvard, 1970), 1–59.

[5]Given the scope of this introduction there is no space for a detailed discussion of rights. In many of the papers included here, the concept of having rights plays a major part in the human/nonhuman, person/nonperson distinctions.

[6]Richard Rorty, *Philosophy and the Mirror of Nature* (Princeton, 1979). Especially see Chapter 2, "Persons Without Minds".

[7]Michael Tooley, Michael has written a number of important articles on the concept of personhood, one of which appears in this volume. His recent book, *Abortion and Infanticide* (Oxford, 1983), contains an interesting discussion of rights and persons.

[8]Michael Tooley, *Ibid,* and Peter Singer, *Practical Ethics* (Cambridge, 1979), and in other works, are leaders in this movement away from the traditional conception/scope of the person.

[9]New Jersey Supreme Court, "In the matter of Karen Quinlan, an Alleged Incompetent," Court opinion of Chief Justice Hughes, reprinted, in part, *Contemporary Issues in Bioethics,* 2nd edition, Tom L. Beauchamp and LeRoy Walters, eds. (Wadsworth, 1982).

[10]In calling these papers "introductory," I do not mean to imply that they are either simple minded or unimportant. They are neither. In fact, Becker's paper at times requires much concentration for understanding. I am reminded of Benson Mates' excellent book *Elementary Logic*, which is anything but elementary.

[11]Page 37 in this volume.

[12]In Michael Tooley's influential "Abortion and Infanticide" (*Philosophy & Public Affairs*, **2**, 1, 1972), we have a case of a single attribute being considered as sufficient for possession of a "serious moral right to life." The attribute is self-consciousness. Harry Frankfurt opts for "second-order volitions" as sufficient for personhood. *See* essays 3 and 5 in this volume.

[13]Page 75 in this volume.

[14]Joel Feinberg, "Abortion," *Matters of Life and Death*, Tom Regan, ed. (Random House, 1980), p. 133.

[15]Page 48 in this volume.

[16]Page 87 in this volume.

[17]Tooley has some further interesting things to say about various potentiality principles in Section 3 of the present paper and in *Abortion and Infanticide*, op. cit.

[18]Michael Tooley, "A Defense of Abortion and Infanticide," *The Problem of Abortion*, Joel Feinberg, ed. (Wadsworth, 1973).

[19]*See* Mary Anne Warren's "On the Moral and Legal Status of Abortion," *Monist*, **57**, 1, January 1973, 43–61.

[20]Page 124 in this volume.

[21]Page 146 in this volume. Though Dennett does not go on to present a detailed account of this distinction, there are references to it throughtout the essay. He does finally conclude that the two senses of person are not separate and distinct. Rather, they are "unstable resting points on the same continuum."

[22]Page 146 in this volume.

[23]Page 175 in this volume. Note how Dennett's moral and metaphysical senses collapse into the "strict" sense here.

[24]*See* Noonan's "An almost Absolute Value in History," op. cit., pp. 34 ff.

[25]Page 253 in this volume.

[26]Robert Joyce argues that the human sperm and ovum are not possible persons because at conception a new being is created, with the sperm and egg being destroyed. *See* Section 2 of Joyce's essay in this volume.

[27]Pages 281–282 in this volume.

The "Beginning" of Human Life

Philosophical Considerations

Daniel Callahan

The Question May Be Placed: When does human life begin? But what is being asked when that question is raised? Is a point in time being sought, some moment in gestation where a line is crossed that differentiates the human from the nonhuman? If a point in time is the aim, for what purpose is it sought: to begin a chapter in an embryology textbook, or to fashion a law dealing with the disposal of embryonic or fetal remains, or to solve an abortion problem? The aim we have in mind in placing the question in the first place will have a bearing on the answer we consider appropriate. Then there are further distinctions to be made. If a moment in time could be specified—"at *this* moment human life begins"—does this entail that the life so begun has or gains value at that moment? Or would it be possible to say, on the one hand, that life begins at "x" moment, but on the other, need not be valued until "y" moment? Again, for what purpose are we asking the question "When does human life begin?" Is our purpose descriptive or classificatory? Or is our purpose legal, social, or moral? People do not usually ask such questions out of idle curiosity. They ask them because they arise in the context of some particular problem; and that context will normally have much to do with what they count as an appropriate answer. In this book, the context of the question is the problem of abortion; the answer given must be helpful in that context, appropriate to the type of concern at issue.

That is easier said than done. In the setting of abortion problems, no consensus whatever seems to exist on the beginning of human life. There are those who think it begins with conception and those who think it begins later. Among the latter, no consensus exists about when that "later" is. While the differences in opinion are well known, there have been few systematic attempts to investigate these differences, seeking some sociological patterns. The only studies worth citing are those which were undertaken by Andie L. Knutson of the University of California, Berkeley, in the mid-1960s. In a 1965 survey of 76 student public-health workers (56 Americans, 20 from other countries), he found a wide range of opinion. One interviewee believed that human life began before conception; 27, at conception; 2, during the first trimester; 8, during the second to third trimester; 10, at birth; 13, at viable birth; 13, sometime after birth (with no answer being given in two cases).[1] While he found that these differences of opinion cut across all sex and religious lines (with differences within every religious group), some tendencies were discernible:

> As expected, those who employ spiritual or religious definitions of when life begins] tend to place the beginning of a life earlier than those who employ psychological, sociological, or cultural definitions. Those who refer to the biological growth process tend also to define a human life as beginning at conception or sometime prior to birth.[2]

Women, he found, tended "to believe that a human life begins at earlier point in development...[and] tend to focus their definitions in terms of biological growth processes with greater frequency than men who, on the other hand, were alone in mentioning independence from the mother, personality, or socio-cultural definitions."[3] He concludes his study by nothing that the wide range of personal beliefs is often not consonant "with the definitions assumed or stated in medical, hospital, or legal codes...The situation tends to foster many types of personal and professional conflicts relating to beliefs, values, and professional responsibility."[4] A later, more extended survey achieved substantially the same results.[5] However, Knutson, in the later study, also tried to determine the felt relationship between the beginning of human life and the value to be assigned to that life. He found that "the assignment of full value appears to be a variable that is to a good extent independent of definitions of a new human life. Some persons define a new life as human long before they assign full value to it; others assign full value to it before they define it as human."[6]

Though Knutson notes that his samples were very small and, of course, restricted to public health workers, it seems unlikely that a much larger sample would achieve substantially different results. And one might guess that a survey taken of the general public or of medical and biological specialists would provide few surprises. People differ on the issue; that much even the casual observer can perceive. In any event, philosophically taken, a survey of beliefs and opinions would hardly provide an answer to the question; it would only establish what people *believe* to be the answer (which is, naturally, important for political and legal purposes).

Obviously people bring to the question different backgrounds and different heritages, not to mention their own personal way of looking at the world. In most circumstances, that is enough to account for differences of opinion. But in this case, I believe, something more accounts for the differences. It is simply a very hard question, and there are no self-evident ways of finding an answer. The biological facts may be evident enough, but—I will try to argue—these facts are open to a variety of interpretations, no one of which is undeniably entailed by the facts. When one moves to the even more difficult question of when value should first be assigned to human life, the range of possible interpretations is just as great. Not only do people differ about the answer to the question, they also differ about how they ought to go about looking for an answer. There exist no universally accepted religious or scientific rules for handling a problem *of this kind*. Every answer will presuppose a different way of looking at the world, at least to some degree. Pluralistic societies are noteworthy precisely because their citizens do not share a common way of looking at the world. An agreement on "the facts" by no means ensures an agreement on the meaning and moral implications of the facts.

How ought the question to be approached? It is best approached by stages, in order to bring out the kind of theoretical and methodological problems hidden beneath the bland wording of the question. At each stage a decision must be made, and I will indicate what I take to be a sound decision, but also indicate other possible alternatives. For the most part, however, most of these segmental conclusions will be primarily negative, serving to bracket those conclusions that seem legitimate (within a given range) and those which do not. The first stage raises the problem of choosing a philosophical perspective on biological "facts." The second stage raises the problem of determining where we get our concept of "human" or "human life." The third stage, with these earlier determinations in hand, requires that one look at the particular facts in question, to see how they could or should be interpreted. The fourth stage raises the problem of how we might go about establishing a "moral policy" in light of the (interpreted) facts. At that stage, the problem of assigning a value or values to the interpreted facts must be confronted for that is the essence of formulating a moral policy.

Philosophical Perspectives

I will begin with a basic philosophical assertion, which can be defended only in part here: Biological data, however great the detail and subtlety of scientific investigation, do not carry with them self-evident interpretations. There are no labels pasted by God or nature on zygotes, primitive streaks, or fetuses that say "human" or "nonhuman." Any interpretation of the known facts is going to be a result not only of our particular interests as we go about establishing criteria for interpretation, but also of the kind of language and the type of analytic-conceptual devices we bring to bear to solve the problems we set for ourselves. This is only to say, at the very outset, that a purely "scientific" answer to the question of the beginning of human life is not possible. "Science" itself is a human construct—a set of methods, terms, and perspectives—and any use of science to answer one particular question, particularly when the answer has moral implications, will be a human use, that is, a use subject to human definitions, distinctions, and decisions. The language of science is a human artifact; the word "life" is a word devised by human beings in order to refer to certain phenomena which can be observed in nature. Scientific method can classify and analyze the phenomena and draw certain "scientific" conclusions (e.g., establish empirical correlations, causal relationships, etc.). But the conclusions it draws will be a result of the humanly devised conceptual schemes used to approach the phenomena in the first place.

As Pierre Duhem put it in a classic statement about the inextricable relationship of the scientific theorist to the data he theorizes about, "It is impossible to leave outside the laboratory door the theory that we wish to test, for without theory it is impossible to regulate a single instrument or interpret a single reading."[7] If this is true, say, of the word "gene," it is all the more true of the word "human." The latter is also a word created by human beings to talk about certain phenomena, which are of interest to that biological species which human beings themselves decided to call "human beings." The problem of specifying a given point in time as critical for the development or emergence of human life is also going to depend on our purposes in looking for such a point. Embryological facts do not shout at us, "Draw a line here!" or "Draw a line there!" Well-developed adults will draw the lines and affix the labels. And they will draw them in different places and affix the labels at different stages, which they do.

There are few scientific arguments about the broad outline of what is going on at different stages of embryological or fetal development. For all that, men who can agree on the biological facts can and do differ when it comes to saying that certain embryological facts *prove* the presence of a "human being." It is neither plausible nor reasonable to (a) assume that one group of scientists, theologians, or philosophers understands the "facts" better that another (for the "facts" are not all that obscure, open only to "correct" interpretations by a gifted handful); or (b) assume that some future scientific discoveries will decisively answer the question about when human life begins; or (c) expect that, with enough scientific "objectivity," a consensus on the "meaning" of the facts could be established for the purpose of ethical discourse on abortion or any other moral problem. To ask people simply to "stick to the facts" is naïve. The "facts" must be used and interpreted, and science provides no fixed rules for the interpretation of facts in moral reasoning. Once this is seen—and the point, for all its simplicity, is crucial—it will be evident that people not only have the right, but are forced to bring extrascientific values and conceptual systems to bear on the facts. As James M. Gustafson has pointed out, "The *values* of human life have not appeared more clearly, because we have a more accurate account of the *facts* of life."[8]

Michael Polanyi and Marjorie Grene have addressed themselves pertinently to some underlying philosophical issues here. In *Personal Knowledge*, Polanyi wrote:

> Our most deeply ingrained convictions are determined by the id-
> iom in which we interpret our experience and in terms of which we
> erect our articulate systems. Our formally declared beliefs can be
> held to be true in the last resort only because of our logically ante-
> rior acceptance of a particular set of terms, from which all our ref-
> erences to reality are constructed.[9]

In short, the way we interpret reality will depend upon our prior
choice of some particular set of terms with which to do the interpreting. Before we can deal with "facts" at all, we need an "idiom," a particular way of going at the facts. Marjorie Grene has
pushed the same point a step further: "Even for apples and pop
songs, let alone for human lives, criteria do not present themselves on the face of things for which they are criteria, but have to
be discovered—or decided?—in the light of some standard, to
which we voluntarily submit ourselves as the right standard for
judging this particular kind of thing."[10] To this point, Miss Grene
adds still another, arguing that "even the least evaluative, most
'factual' judgments depend for the possiblity of their existence on
some prior evaluative act."[11] This is only to say that what we
choose to call a "fact" will be dependent, at root, upon some evaluative system which enables us to distinguish between a "fact"
and a "nonfact"; and it goes without saying that this evaluative
system will also be a human creation. This is not to deny there is
a reality outside of our language and conceptual systems. It is
only to affirm that we cannot get at this reality or interpret it without words and concepts and invented methods.

   Both Polanyi and Grene share an emphasis, therefore, on prior,
anterior acts of evaluation; we approach facts with these evaluations in hand. But what kind of prior evaluative acts ought to be
brought to bear on the question of when life begins? Polanyi and
Grene do not address themselves to this question (or to the abortion problem), but both provide a useful framework for approaching such a question. The very fact that we ask the question at all—
using words like "human" and "life"—shows that in some sense
(to be determined) we already possess, as shown in our use of language, some prior concepts; our words did not appear out of no-
where. When we ask the question about the beginning of human
life, therefore, it is to be supposed that we already have some
more or less determinate meaning in mind for these words. Our

problem is to relate these words—and the corresponding concepts—to the available biological data.

But an acknowledgement seems necessary before we do so. We should recognize that we already know the *outcome* of the human gestational process; we know "where babies come from." It is because we know this process that we are able to work our way back to seek its beginning; our prior anterior knowledge has enabled us to know that, e.g., zygotes found in human females are *human* zygotes (which give place to human embryos, fetuses and neonates). In Polanyi's more formal terms:

> The analysis of the process by which living beings are formed corresponds to the logic of achievement, as illustrated by the manner in which we find out how a machine works. We start from some anterior knowledge of the system's total performance and take the system apart with a view to discovering how each part functions in conjunction with the other parts. The framework of any such analysis is logically fixed by the problem which evoked it. Its contents may be extended indefinitely and it may penetrate thereby even further into the physical and chemical mechanism of morphogenesis; but its meaning will always lie in its bearing on living situations that are true to type, emerging from a mosaic of morphogenetic fields.[12]

In our case, abortion, the "framework" of which Polanyi speaks has been fixed by our anterior knowledge of postnatal human beings. We know generally what a postnatal human being is and our problem is to find the point when it can be said this "human being" came into existence as a human being. But we could not undertake such an analysis in the first place did we not begin with a prior knowledge of what a human being (once born) is.

To talk in this way is to talk holistically and teleologically. It is to say that we cannot answer the question we have set unless we begin our analysis with postnatal human beings and then work backwards into the gestational process, observing the adult-aimed, forward-moving stages of development. Marjorie Grene has sketched the philosophical presuppositions in such an approach:

> What the higher level of organization does is in some sense to *control* the lower. Parts become the parts they are in relation to the whole, organs are organs *as* their function dictates, embryos de-

velop toward their specific norm....True, organic life cannot exist
without both levels: wholes without their parts, functions without
the mechanisms needed to perform them, or living individuals
without the whole development from fertilized eggs which produced them. But in some sense the higher level provides a principle which orders or determines the lower. In every case the lower
level specifies conditions, while the higher gives us principles of
organization, ends or reasons. The conditions are indeed necessary, but while it is possible to understand the principle without
reference to the detailed conditions, it is *not* possible to understand
the conditions *except* as conditions of the whole, the activity, or the
endpoint of development on which they bear.[13]

Later she asked: "Can the biologist preceed to describe what it
is he is analyzing without referring to structures, uses or achievements? I think not. In other words, teleological discourse has not
only a regulative but at least a *descriptive* function within biological research."[14]

As an alternative to the use of teleological language, the psy-
chologist Edna Heidbreder employs the concept of *Gestalten:*

> *Gestalten* are found in processes outside the psychological field
> altogether. In biology, the process of ontogenesis shows striking
> examples. Here the orderly development of definite organic struc-
> tures from the primitive germ-layers gives both spatial and tempo-
> ral *Gestalten* remarkably similar to those found in psychology.
> They show the same orderly sequence, the same continuity, the
> same progress toward a given end, the same relation of particular
> processes to the whole in which they are involved.[15]

To be sure, not all biologists or philosophers are by any means
happy with teleological analyses.[16] But this is not the place to
argue out all the myriad philosophical and scientific issues involved in the legitimacy of such an analysis, which would take us
far afield. Suffice it to say that, in this context, such an analysis
appears extremely helpful. It points out the necessity, if we are to
understand when human life "begins," of starting from postnatal,
developed human life, seeking its origin and those stages of development which lead to an ultimate full human development. To
those who distrust this method of procedure it is open to propose
an alternative philosophy of biological analysis, suitable to the
present problem. The only thing that seems to me wholly unac-

ceptable here is to ignore altogether the need to provide and explicate *some* philosophy of biological analysis. Unfortunately, that need is often ignored in abortion discussions.

The philosophical perspective I am going to employ and urge is teleological. In a word, if we are to make sense of zygotes, embryos, and fetuses, we cannot *begin* our analysis with these entities. Prior to investigating them, we must investigate what the phrase "human life" means or can mean; and this means beginning our analysis with fully developed human beings, working back from them to the initiating stages of human development. Whatever "human life" or "human being" means, it is safe to say that these concepts have not come into use by virtue of an analysis of exclusively prenatal life. Instead, they have been developed by an analysis of postnatal developed human life; and it is this analysis which must be employed, explicitly, in trying to answer the question "When does human life begin?" The only warrant for calling a given zygote a "human zygote" is that a prior knowledge exists of the developmental process whereby those zygotes found in pregnant human females give rise eventually to developed "human beings."

In order, then, to answer our question, it is first necessary to examine the concept "human." This means beginning with the concept "human" as it is employed in describing developed human beings. That done, we will be in a position to examine the prenatal biological data, to see how the concept of "human" can or ought to be applied to that data.

The Meaning and Use of "Human"

To ask what the word "human" means is to ask about the nature of man. What is "man" and what is "human nature"? To be "human" means to possess those characteristics we associate with the word "man." The problem of man and human nature is ancient; all religions and seemingly all important philosphies have tried to provide an answer. Of late, a spate of books and symposia have appeared on the topic, a symptom of its renewed importance in an era of social and technological change, and a symptom as well of considerable uncertainty.[17]

For the most part, historical efforts to define the nature of man have taken the form of a search for one overriding characteristic that constitutes man's essence. Boethius' classic definition of a person—*"Persona est substantia individua rationalis naturae"*— reflects his work as a translator of Aristotle and, more broadly, the Greek tradition, which saw in rationality man's essence: man is a rational animal. This tradition was carried through the ages, manifesting itself in medieval philosophy, the Renaissance, and into the Enlightenment. Though later philosophies and cultural currents located the essence of men elsewhere—*zoon politicon* (man as social being), *homo faber* (man as producer), man as symbolmaker—the search for an essence has continued down to the present. Thus Stephen Toulmin has written that "the process of intellectual growth is the salient point in all human development in the development of society, in the development of the individual in economic development, and in cultural development alike."[18] Nathan Scott has written of man as that being which is "open to the ineffable."[19] Willard Libby has affirmed that "this, to me, is man's place in the physical universe: to be its king through the power he alone possesses: the Principle of Intelligence."[20]

It is unnecessary to multiply examples of efforts to locate an essence of man; they are familiar enough.[21] Historically, they have been closely tied to attempts to distinguish men from animals, a practice which also continues into the present. "Man is unique," Ernst Mayr has written. "He differs from other animals in numerous properties such as speech, tradition, culture, and an enormously extended period of growth and parental care."[22] The great anthropologist A. L. Kroeber wrote:

> Man is an essentially unique animal in that he possesses speech faculty and the faculty of symbolizing, abstracting, or generalizing. Through these two associated faculties he is able to communicate his acquired learning, his knowledge and accomplishments....This special faculty is what was meant when someone called man the "time-binding" animal. He "binds" time by transcending it, through influencing other generations by his actions.[23]

Marjorie Grene argues that men can be distinguished from animals in that the former are capable of thinking, speaking, knowing, achieving, and community-making.[24] For George Herbert

Mead the difference lies in man's unique ability to use symbols and to distinguish symbols from objects.[25] Ernst Caspari, surveying the evolutionary evidence, has said that "that differences in mental abilities between man and his nearest relatives are very large. It is impossible to enumerate them....Suffice it to point to the increased ability for and dependence on learning in all behavioral activities, the ability to communicate by speech, and the ability to make tools."[26] A. Irving Hallowell has said that "it seems rather to be the manner in which experience is organized that sets the human line apart from other animals."[27]

Again, though, it is not necessary to multiply examples. More important for our purposes are some of the methodological problems. *For the purposes of an abortion discussion,* what kind of an analysis of the concepts "human," "man," "human nature" is needed? Is it imperative that we be able to stake out and define the "essence" of the "human," or will some other kind of description do? If we must, of necessity, work backwards through the gestational process, using as our starting point the developed human being, what *kind* of a concept of "human" are we looking for? This seems to me an extraordinarily difficult question to answer in our context. This much at least can now be said: We need a concept sensitive to the findings of a number of disciplines. Fortunately, there exist a variety of attempts to find usable definitions of "human," attempts which, precisely because they are far removed from abortion polemics, are especially valuable. We can, I think, try to learn something from these efforts; they provide some suggestive hints.

In zoology, attempts to grapple with the "species problem" provide one example. In essence the problem is this: what is a species, what are the criteria for including a being in a given species, and what distinguishes one species from another? At the level of taxonomy, it is a problem of classifying different kinds of living things. At another level, it is the problem of determining the stages in human evolution (when did "human beings" first appear?). The most helpful discussion of this problem remains a book edited by Ernst Mayr, *The Species Problem.*[28] In his own essay, "Species Concepts and Definitions," Mayr notes that the history of the problem has included all shades of opinion, ranging from the view that species exist in nature independent of human classification to the opinion that species are a human invention, without a clear referent in nature. Among zoologists, two leading schools of opinion have contended with each other, the typological (using essentially morphological criteria) and the populationist (using essentially biological-genetic criteria). J. Imbrie distinguishes the two schools in the following way, according to their leading concepts: "The *typological* concept defines species as a group of individuals essentially indistinguishable from some specimen selected as a standard of reference. The *biological* species concept...considers the species to be made up of one or more inbreeding populations."[29] Philosophically put, the typological concept (according to Mayr) is Platonic: it begins with the assumption that each species manifests an "idea," some common, underlying trait which a given specimen manifests and which is shared by other specimens of the same species: "the typological species concept treats species merely as random aggregates of individuals which have the 'essential properties' of the 'type' of species."[30] The problem which Mayr and others discerned is that typological concepts are "static," having no conceptual place for individual variations:

> The assumptions of population thinking are diametrically opposed to those of the typologist. The populationist stresses the uniqueness of everything in the organic world. What is true of the human species, that no two individuals are alike, is equally true for all other species of animals and plants....For the typologist, the type (*eidos*) is real and the variation an illusion, while for the populationist the type (average) is an abstraction and only the variation is real.[31]

One consequence of an employment of the concept of a "population" has been the demise of "single-character taxonomy." As Carelton S. Coon has written: "[The] obsolete concept of single-character taxonomy has long since been abandoned. Zoologists now base their decisions on all the characteristics they can identify and measure, characteristics which together give the animal its essential nature, its (to borrow a psychological term) *gestalt*."[32]

An alternative to single-character taxonomy is that of a "population" concept, which stresses all the variations within a species, but particularly genetic variations. It is part of what George Gaylord Simpson (among others) called the "new systematics," a term meant to signal (in the late 1930s and early 1940s) the shift from a reliance upon morphotype (a single morphological characteristic) to that of a population, which is concerned with "*all* of its members collectively, with their resemblances and differences."[33]

As part of this argument, J. M. Thoday has stressed the importance of genetic considerations:

> Natural populations are of fantastic genetic complexity, and natural environments are also complex....It is becoming more and more clear, the more experiments we do on the genetics of natural populations, that the old idea, itself relating to the biological type concept, that populations are genetically rather uniform, there being by and large one normal or wild-type, along with many abnormalities, each rare, is totally misleading. Normal flies or normal men, comprise an extensive array of different genotypes....So extensive is their variety that we may say without exaggeration that, apart from identical twins, no two individuals are, or ever have been, genetically exactly alike.[34]

"There are," he adds, "as many human natures as men," a proposition which has also been affirmed by T. Dobzhansky: "The nature of man as a species resolves itself into a great multitude of human natures."[35]

At this point, however, we run into an obvious philosophical problem, one familiar to any student of the age-old problem of "universals." If it is impossible, or at least unfruitful zoologically, to use any kind of "single-character" specification of a species, what warrant is there for talking of a "human species" at all? Or, to put it another way, what warrant is there for calling "x population" in nature a "human population"? Can one speak, as Dobzhansky and Thoday do, of "many human natures," each different, without rendering the word "human" meaningless? Not really if one takes the "many" literally, for unless there is some tacit, generalizable understanding of what the word "human" means, some universal signification, then it could not be used to describe more than one organic entity. One could not even use a phrase like "a great multitude of human natures" unless the word "human" was meant to convey something about the multitudes as a particular multitude. They are a multitude of "*human* natures," not a multitude of some other kind of natures. As Marjorie Grene has observed: "We can only make sense of experience through its subsumption under universals, and that such universals act as standards for the evaluation of experience, or rather that we act in submission to these standards as judges of experience: this much in the Platonic account we must admit."[36] George Gaylord Simpson almost, but not quite, saw the philosophical problem:

> It is a convenience to a systematist engaged in identification and cataloging to recognize taxonomic units by the characters common to all their members. Such a procedure is, indeed, virtually necessary in practice, but it is inherited from the old systematics and lends itself to serious philosphical confusion. The characters-in-common may become a morphotype in the mind of the classifier. He tends to think a category is defined by these characters.[37]

What Simpson might have asked himself is why such a procedure is "virtually necessary in practice." For one cannot begin identifying and cataloging *at all* unless one possesses some notion, however inchoate, of the common characteristics of a population. This is the kind of point Miss Grene is trying to make in attempting to show the futility of a rigidly nominalist approach to biological entities. This is not to argue, though, that a "single-character taxonomy" is again called for; it is only to argue that, in talking about what counts as "human," the term must have some meaning that will cover a "multitude" of instances.

What we can profitably learn from the struggle over the "species problem" is twofold. First, a useful definition of "human" must be one which takes account of the widest possible diversity of instances, an entire "population," and which implies a consequent skepticism toward single-character or essentialistic definitions; second, a recognition that genetic criteria are more serviceable than morphological criteria in distinguishing a species. On the second point, Mayr argues:

> The very fact that species is a gene pool...is responsible for the morphological distinctness of species as a by-product of their biological uniqueness. The empirical observation that a certain amount of morphological difference between two populations is normally correlated with a given amount of genetic difference is undoubtedly correct. Yet, it must be kept in mind at all times that the biological distinctness is primary and the morphological differences secondary.[38]

The significance of this conclusion for a discussion of abortion, and concretely in order to answer our question of when human life begins, is this: when we talk of "human life" we must try to take account of a wide variety of "human characteristics," and we must also, in the process, not fail to take account of the genetic characteristics of human populations and human life. The import of this conclusion will be developed shortly, but first another example is in order, this time drawn from the field of anthropology.

One of the key problems of anthropological theory is that of the relationship between human biological nature and human social culture. As Marvin Harris has written,

> One of the basic requirements of a theory of cultural continuity and change is a description of what used to be called human nature. All cultural items are partially the product of a set of biophysical constants which are shared by most if not all *Homo sapiens.* Correct though it may be that the explanation of differences and similarities cannot be achieved merely be invoking these constants, it is no less improbable that explanations can be achieved without them.[39]

But the problem of finding the exact relationship between the biological and the cultural has proved vexing. On the one hand, there are those who would stress culture as the sole determinant of human behavior, excluding the biological altogether.[40] On the other are those who have seen everything as biologically determined.[41] Others, however, have tried to find a synthesis of biological and cultural determinants. A. Irving Hallowell, for instance, has pointed out that psychoanalytic theory has helped to clarify (though hardly altogether) those "specific differences in personality structure and functioning which can be shown to be related to cultural differences."[42] He quickly adds, though:

> Implicit in these data are indications that universal dynamic processes are involved which are related to the psychobiological nature of modern man as a species. Likewise, capacities are implied which must be related to generic psychological attributes of *Homo sapiens* that have deeper roots in the evolutionary process.[43]

Hallowell's particular concern is that an emphasis on the cultural characteristics of man not be allowed to open an unbridgeable gap between man and the other primates, which he believes would obscure the evolutionary problem.

One thinks of such characterizations of man as "the rational animal," the "tool-making animal," the "cooking animal," the "laugh-
ing animal," the "animal who makes pictures," or *animal symboli-
cus*. All these characterizations stress man's differences from other
living creatures. Like the criteria of culture and speech, they
emphasize discontinuity rather than the continuity, which is likewise inherent in the evolutionary process.[44]

His aim is to stress the importance of biological, psychological
and cultural interrelationships:

Although no unanimity of opinion has been reached, hypotheses
should emerge in time which will lead to further clarification of the
relations between neurological evolution, psychological functioning and cultural adaptation. Of central importance in this complex
web of relationships is the distinctive psychological focus in *Homo
sapiens*—the capacity for self-objectification which is so intimately linked with the normative orientation of all human societies.[45]

Thus, while "the capacity for self-objectification" is, for Hallowell, the salient feature of *Homo sapiens,* this capacity must be
understood in biological as well as cultural terms, with the psychological forming a major connecting link.

Clifford Geertz is another anthropologist who believes that a
synthesis of biological and the cultural is needed, though he
comes down much harder on the cultural side than Hallowell:

Whatever else modern anthropology asserts, it is firm in the conviction that men unmodified by the customs of particular places do
not in fact exist, have never existed, and most important, could not
in the very nature of the case exist....This circumstance makes the
drawing of a line between what is natural, universal, and constant
in man and what is conventional, local, and variable extraordinarily diffcult. In fact, it suggests that to draw such a line is to falsify
the human situation, or at least to misrender it seriously.[46]

In another article, he brings his position to a fine point;

The apparent fact that the final stages of the biological evolution of
man occurred after the initial stages of the growth of culture
implies...that "basic," "pure," or "unconditioned" human nature, in
the sense of the innate constitution of man, is so functionally
incomplete as to be unworkable. Tools, hunting, family organi-

zation, and later, art, religion, and a primitive form of "science," molded man somatically, and they are therefore necessary not merely to his survival but to his existential realization. It is true that without men there would be no cultural forms. But it is also true that without cultural forms there would be no men.[47]

Geertz contends that anthropological theory, in its attempt to achieve "an exacter image of man," needs unitary systems of analysis, wherein the biological, psychological, sociological and cultural factors can be treated as variables within the systems. At present, he notes, these variables are "sequestered in separate fields of study," making it difficult to develop unitary anthropological systems.[48]

Man [he says] is to be defined neither by his innate capacities alone...nor by his actual behavior alone, as much of contemporary social science seeks to do, but rather by the link between them, by the way in which the first is transformed into the second, his generic potentialities focused into his specific performances. It is in man's *career*, in its characteristic course, that we can discern, however dimly, his nature.[49]

As an example of a synthetic statement, illustrating a thrust toward a unitary theory, Geertz writes: "As our central nervous system—and most particularly its crowning curse and glory, the neocortex—grew up in great part in interaction with culture, it is incapable of directing our behavior or organizing our experience without the guidance provided by systems of significant symbols."[50] For Geertz, then, it is culture which enables man to develop his innate capacities. At the same time, his innate capacities themselves have been in part determined by culture—it is a two-way process. Most importantly,

...the extreme generality, diffuseness, and variability of man's innate (i.e., genetically programmed) response capacities means that, without the assistance of cultural patterns he would be functionally incomplete...a kind of formless monster with neither sense of direction nor power of self-control, a choas of spasmodic impulses and vague emotions.[51]

It is evident, of course, that I am here using only a few anthropologists to bring out some of the problems. But this is not the place for a survey of anthropology. It is sufficient to say that the

problems the cited authors are circling—the relationship between
biology (normally thought of as determining what is "innate") and
culture (normally thought of as determining what is "conditioned")—have a considerable bearing on our particular problem.
For when we ask what "human" means, we should be asking—if
we have learned anything from anthropology—about the relationship of biology and culture, recognizing the importance of culture
in contributing to the formation of a "human being." The late psy-
chologist Gordon Allport seems to me to have illuminated a related problem. His concern was to show the weaknesses of general
(i.e., typological) categories in explaining the human process of
"becoming" and individual uniqueness. His way of working out a
synthesis of biology, psychology and culture is well expressed in
the following lines:

> Each person is an idiom unto himself, an apparent violation of the
> syntax of the species. An idiom develops in its own peculiar con-
> text, and this context must be understood in order to comprehend
> the idiom. Yet at the same time, idioms are not entirely lawless and
> arbitary; indeed they can be known for what they are only be com-
> paring them with the syntax of the species.[52]

In sum, consistent with major thrusts of zoology, anthropology,
and psychology, any definition of the "human" must take account
of the interaction of biological, psychological, and cultural factors; all have their place and none is dispensable. In terms of the
abortion problem, one can thus say that any definition of "human"
in the question set for this chapter will be defective if it ignores the
interrelationship of the three mentioned factors. More pointedly,
"human" cannot be defined in a genetic way *only*, or a psychological way *only,* or a cultural way *only;* it must be defined in such a
way as to take account of all three elements in the "human."
   If it is possible to discern in the biological and social sciences a
distinct movement in the direction of enriched, comprehensive
definitions of man—an antimonotypism—the same trend is discernible in the social sciences and in theology. Some examples
must suffice here. The Jewish philosopher Martin Buber, for instance, personified (and personally stimulated) two major elements in contemporary theological anthropology: a stress on the
full complexity of the human and on the centrality of human rela-

tionships. "A legitimate philosophical anthropology," he wrote, "must know that there is not merely a human species but also peoples, not merely a human soul but also types and characters, not merely a human life but also stages in life."[53] The extent to which this statement echoes a number of the quotations above does not need underscoring. Resistant to any one-dimensional definition of man (for instance, man as a "rational animal"), Buber saw the essence of man in his relationship with other men (which presupposes rationality, affectivity and the power of willing): "The philosophical science of man, which includes anthropology and sociology, must take as its starting-point the consideration of this subject, 'man with man.' If you consider the individual by himself, then you see of man just as much as you see of the moon; only man with man provides a full image."[54] It is man's capacity to enter into relationship which is his distinctive attribute. A similar point has been made by an English Dominican:

> To discover what it is to be human and to achieve properly human fulfillment account must be taken of man not simply as a biological object, not even simply biologically (which would introduce a whole consideration of his ecology), but in the specifically human dimension in which he enters into communication with others at a human level. That is "natural" to man which constitutes him not merely in isolation, but in relation to the whole world-for-man which he creates around him...and in relation to other persons who stand not simply as objects but as other subjects around him.[55]

Ernest Becker, from a nonreligious perspective, has pressed on the social sciences the importance of a cross-disciplinary, unified concept of man:

> In the human sciences man must be seen at all times in the total social-cultural-historical context, precisely because it is this that forms his "self" or his nature...The narrow positivist fallacy has always been that one can somehow know the object in itself, that it exists as a thing in nature and must be isolated and defined. But this is a species of essentialism...We come to know a thing, furthermore, only in terms of its relationships, never in itself.[56]

Finally, one can cite the eminently judicious words of Erich Fromm and Ramon Xirau, in summing up their excellent examination of the historical course of discussions on the nature of man:

It can be stated that there is a significant consensus among those
who have examined the nature of man. It is believed that man has
to be looked upon in all his concreteness as a physical being placed
in a specific psychical and social world with all the limitations and
weaknesses that follow from this aspect of his existence. At the
same time he is the only creature in whom life has become aware of
itself, who has an ever-increasing awareness of himself and the
world around him.[57]

The Importance of "Potentialities"

The stress on comprehensiveness, richness and flexibility in
defining the "human," common to all the authors so far cited,
stems from one overriding consideration. When human beings
are looked at in all their diversity, as individuals or as members of
a species, it becomes extraordinarily difficult to single out one
human attribute that can be counted as normative and decisive.
Human beings are rational and irrational, individual and com-
munal, biological and cultural. Any single definition, stressing
one attribute, invariably fails to catch the full measure of man. If
this is true when one tries to develop a concept of the "human"
using developed, adult humans as evidence, it becomes all the
truer when one begins dealing with the borderline cases or individuals who display some human characteristics but not others.

Hence, it becomes imperative that any attempt to define "human life" take account of "potentialities" or "capacities." To be "human" is not just to display, here and now, the full range of human
characteristics. Geertz, one recalls, stressed that the whole career
of man must be taken into account in defining him. But that career is especially noteworthy precisely because, at any given moment in life, a human being may be displaying very few human
characteristics. Human beings do not always think; sometimes
they are asleep, or drugged, or too young to think. Human beings
are not always in relationship with others; sometimes they are
alone. It is, in fact, quite easy to imagine a great variety of situations where a whole range of human potentialities may not be
realized presently. Yet a mere lack of present actualization of
human potentiality would not be sufficient warrant for withdraw-
ing the appellative "human" from a being morphologically or
genetically human.

One way or the other, we are forced to ask, in cases where actualized human characteristics are not displayed, about the potentiality for such characteristics appearing. The zoological concept of a "population" presupposes the use of "potentiality" as part of any criterion of inclusion in a population: the potentiality for interbreeding. An anthropological concept of man, laying a stress on man as culture-maker, will have to take account of the human potentiality for culture-making. And, as Allport has argued, an understanding of the psychological process of "becoming" must likewise make use of concepts of "potentialities" and "capacities."

Most critically for our purposes, any adequate definition of "human life"—to be consistent with both practice and theory in a variety of disciplines—must include a place for three kinds of potentialities, all interrelated: biological, psychological, and cultural potentialities. Human cultural potentialities cannot be realized without the actualization of biological potentialities, just as psychological potentialities cannot be realized without the actualization of cultural potentialities. To repeat Geertz, man is to be defined by the way his innate (biological) capacities are transformed through the medium of culture into actual human behavior. The complaint of those anthropologists who oppose an exclusively culture-oriented definition of the human is to be heeded. To ignore the biological base of human culture is to leave little or no room for an explanation of the process by which human beings, ontogenetically or phylogenetically, have been able to develop cultures. Marjorie Grene has succinctly summarized the thought of the zoologist Adolf Portmann on the way biological potentialities are transformed into cultural actualities:

> Our unique pattern of development [as human beings] is not an "after thought" tacked onto a standard embryogenesis. The human attitudes and endowments which we must acquire in infancy are prepared for very early indeed in embryonic growth: thus the first preparation for the upright posture, in the development of the pelvis, occurs in the second month of the foetus's growth. The preparation for the acquisition of speech, moreover, involves glottal structures very strikingly and thoroughly different from those of any other species. And the huge size of our infants relatively to the young of apes—born more "mature" but very much smaller—is probably related, Portmann conjectures, to the immense develop-

ment of the brain necessary for the achievement of human rationality—a development which begins, again, very early in ontogene-
sis. In short, the whole biological development of a typical mammal has been rewritten in our case in a new key: the whole structure of the embryo, the whole rhythm of growth, is directed, from
first to last, to the emergence of a culture-dwelling animal.[58]

This last passage, with its key word "directed," reintroduces the
teleological problem mentioned earlier. My working assumption
is that, when we speak of "human life," we must also speak of (*a*)
human potentialities, and (*b*) potentialities in a direction, i.e., not
random potentialities, but potentialities which, speaking teleologically, can (in Grene's and Portmann's sense) be said to be "directed." This term does not imply a "director," but is only meant to
be a way of interpreting the apparent fact that human development
shows certain characteristic patterns and directions. These patterns and directions are *toward* rationality, individuality, culture-
making, language, human relationships, tool-making, and so on.
It is the potentiality of certain kinds of organic beings *toward*
these characteristics that provides us with a warrant for calling
these beings "human." It is because we know the whole human
career (in Geertz's terms) and because we know man as a whole
(to use Polanyi's terms) that we are justified in talking of human
beings in terms of both their actualized realities and their as yet
unrealized potentialities. Our knowledge of the former enables us
to speak of the latter, and to speak of it in teleological language.

It is for essentially the same reasons that one would be justified
in speaking of "levels of organization." If human development,
ontogenetically, represents the unfolding of potentialities, then it
becomes possible to see the process of development as proceeding by and through different stages. A human adult has reached a
different level of organization and complexity than a newborn infant. The problem is how the different levels of organization are
related to each other; in particular, whether and to what extent a
higher, more developed level is dependent upon a lower. The
anthropologist Erich Kahler, discussing A. L. Kroeber's attempt
to distinguish between the physiochemical and the cultural level
in human life, has written:

> *Existence is a multilevel affair.* As a body, I am a natural organiza-
> tion of lower beings, living, moving, changing, growing and de-

caying beings, namely the *cells*. Any change or disturbance in this organization, or even in the organization of the cells themselves, has the most powerful and serious effects on what we may consider the essence or quintessence of the physical system, the *psyche*. This is, after all, recognized in the psychosomatic theory and in recent psychiatry. The psyche, in turn, has a well-established influence on the *mind*....We must also realize that all influence effective between different levels is a two-way process: it works upward as well as downward. There is a mutual interaction going on between mind, psyche, body, and so forth.[59]

Though differing somewhat in their emphasis, both Kroeber and Kahler are doing nothing more than trying to account for an increased awareness of the unity between the biological, the psychological, and the cultural in man; they cannot be explained independently of each other, Marjorie Grene adds another step in the same kind of analysis, adding the note of increasing complexity in the different, higher levels of development: "In proceeding from the recognition of matter to life to persons to responsible persons, we are proceeding up a scale of complexities, each of which entails the earlier levels. Responsible persons are persons, persons are individuals, individuals are physical structures, yet each kind we recognize as also more, and other, than the preceding, or underlying, level."[60] As Polanyi puts it, "In the course of anthropogenesis, individuality develops from beginnings of a purely vegetative character to successive stages of active, perceptive, and eventually responsible, personhood. The phylogenetic emergence is continuous—just as ontogenetic emergence clearly is."[61]

Of course, a biological analysis which speaks in terms of levels and complexities of organization has a long pedigree, going back to Aristotle. But it still remains a serviceable form of analysis; indeed, it is hard to discern an alternative if one wants to know when human life begins. For the very question presupposes that human life can be comprehended in temporal terms—as having a beginning in time and an end in time. There is some reason (as we shall see shortly) for worrying a bit about this presupposition, but no sufficient reason to reject it. And as part of that same presupposition, there is the antecedent knowledge that individual human life does proceed through stages. We do, after all, refer to neonates, infants, children, adolescents, adults, and the elderly. Pediatrics, at one end of the human life—span, and geriatrics, at the other, can exist as independent medical disciplines precisely because there exists a body of knowledge concerning human life at the infant and elderly stages. We know what can or ought to be expected at each stage simply because we know something about the different stages, or levels, of human development. Few would be concerned with genetic abnormalities in the early months of gestation were it not known that these abnormalities have immense consequences for later development. Knowledge of this kind can be termed "teleological knowledge": knowing the direction and likely outcome of a biological process.

A summation is in order before moving to the next stage of this discussion. I have argued, in the first stage, that a teleological analysis of the biological data is legitimate, necessary, and illuminating as a philosophical basis for approaching an answer to the question of when life begins. I have argued, in the second stage, that an important movement in some scientific disciplines concerned with the "human" is that the "human" must be defined not in single-character, or essentialistic, terms but rather in terms of variety and diversity. Moreover, there is considerable agreement that an analysis of the "human" must take account, holistically, of the biological, the psychological, and the cultural; no one of them can be scanted, at the cost of misunderstanding the others. I have also tried to point out the importance of "potentiality" and "capacity" in analyzing "human life," adding an affirmation of the language of "levels of organization" as helpful in understanding the process whereby the potential is made actual.

At this point, I want to draw some conclusions pertinent to a discussion of when human life begins. First, a negative conclusion: any answer to the question which rests on one human characteristic alone is to be suspect. Second, a positive conclusion: any answer to the question must take account of "potentiality," and not rest its case exclusively on achieved human characteristics. Third, a negative conclusion: any answer to the question which ignores the biological, genetic basis of human development and individuality is to be rejected. Fourth, another positive conclusion: the best way to analyze human development in its temporal course is to analyze the different stages of development teleologically, in terms of their direction and potential ends.

Notes

[1]Andie L. Knutson, "When Does a Human Life Begin? Viewpoints of Public Health Professionals," *American Journal of Public Health*, 57 (December 1967), 2167.

[2]*Ibid.*, p. 2169.

[3]*Ibid.*, pp. 2171–2172.

[4]*Ibid.*, p. 2175.

[5]Andie L. Knutson, "The Definition and Value of a New Human Life," *Social Science and Medicine*, 1 (1967), 7–29.

[6]*Ibid.*, p. 26.

[7]Pierre Duhem, "Physical Theory and Experiment," Herbert Feigl and May Brodbeck (eds.), *Readings in the Philosophy of Science* (New York: Appleton-Century-Crofts, 1953), p. 237.

[8]James M. Gustafson, "Christian Humanism and the Human Mind, John D. Roslansky (ed.), *The Human Mind* (New York: North-Holland, 1967), p. 87.

[9]Michael Polanyi, *Personal Knowledge* (New York: Harper Torchbooks, 1964), p. 287.

[10]Marjorie Grene, *The Knower and the Known* (New York: Basic Books, 1966), p. 158.

[11]*Ibid.*, p. 160.

[12]Polanyi, *op. cit.*, p. 357.

[13]Grene, *op. cit.*, p. 233.

[14]*Ibid.*, p. 236.

[15]Edna Heidbreder, *Seven Psychologies* (New York: Appleton-Century-Crofts, 1961), p. 355.

[16]One of the best philosophical discussions of such analyses remains that of Ernest Nagel, "Teleological Explanations and Teleological systems," *Readings in the Philosophy of Science, op. cit.*, pp. 537–558.

[17]*See* especially Erich Fromm and Ramon Xirau (eds.), *The Nature of Man* (New York: Macmillan, 1968); John R. Platt (ed.), *New Views on the Nature of Man* (Chicago: University of Chicago Press, 1965); William Nicholls (ed.), *Conflicting Images of Man* (New York: The Seabury Press, 1966); Paul E. Oehser (ed.), *Knowledge Among Men* (New York: Simon and Schuster, 1966).

[18]Stephen E. Toulmin, "Intellectual Values and the Future," *Knowledge Among Men, op. cit.*, p. 159.

[19]Nathan A. Scott, Jr., "The Christian Understanding of Man," *Conflicting Images of Man, op. cit.*, p. 10.

[20]Willard F. Libby, "Man's Place in the Physical Universe," *New Views on the Nature of Man, op. cit.*, p. 15.

[21]*See* Fromm and Xirau (eds.), *op. cit.*, for a collection of such attempts throughout history.

[22]Ernst Mayr, *Animal Species and Evolution* (Cambridge: Harvard University Press, 1963), p. 623.

[23]A. L. Kroeber, *Anthropology: Culture Patterns and Processes* (New York: Harcourt, Brace and World, 1963), p. 8.

[24]Grene, *op. cit,.* pp. 172ff.

[25]George H. Mead, *Mind, Self and Society* (Chicago: University of Chicago Press, Phoenix Books, 1962), pp. 120–122.

[26]Ernst Caspari, "Selective Forces in the Evolution of Man," M. F. Ashley Montagu (ed.), *Culture: Man's Adaptive Dimension* (New York: Oxford University Press, 1968), p. 165.

[27]A. Irving Hallowell, "Cranial Capacity and the Human Brain," *Culture: Man's Adaptive Dimension, op. cit.,* p. 184.

[28]Ernst Mayr (ed.), *The Species Problem* (Washington, DC: American Association for the Advacement of Science, Publication **50**, 1957).

[29]J. Imbrie, "The Species Problem with Fossil Animals," *The Species Problem, op. cit.,* p. 126.

[30]Mayr, "Species Concepts and Definitions," *The Species Problem, op. cit.,* p. 13.

[31]Mayr, *Animal Species and Evolution, op. cit.,* p. 5.

[32]Carleton S. Coon, *The Origin of Races* (New York: Alfred A. Knopf, 1962), p. 13; Polanyi comments that the downfall of taxonomy lay in its dependence upon "the good taxonomist," with a personal ability to recognize affinity to type (*Personal Knowledge, op. cit.,* p. 351).

[33]George Gaylord Simpson, *The Major Features of Evolution* (New York: Columbia University Press, 1953), pp. 340–341; *see also* J. S. Huxley (ed.), *The New Systematics* (Oxford: Clarendon Press, 1940), and Ernst Mayr, *Systematics and the Origin of Species* (New York: Columbia University Press, 1942).

[34]J. M. Thoday, "Geneticism and Environmentalism," J. E. Meade and A. S. Parker (eds.), *Biological Aspects of Social Problems* (Edinburgh: Oliver Boyd, 1965), p. 101.

[35]Thoday, *ibid.,* p. 103, and Theodosius Dobzhansky, *Heredity and the Nature of Man* (New York: Signet Books, 1966), p. 57.

[36]Grene, *op. cit.,* p. 169.

[37]Simpson, *op. cit.,* p. 341. For a critique of the "population" concept, *see* W. R. Thompson, "The Status of Species," Vincent E. Smith (ed.), *Philosophical Problems in Biology* (New York: St. John's University Press, 1966), pp. 67-126.

[38]Mayr, "Species; Concepts and Definitions," *op. cit.,* p. 13.

[39]Marvin Harris, *The Rise of Anthropological Theory* (New York: Thomas Y. Crowell Co., 1968), p. 429.

[40]*See,* for instance, Benson E. Ginsberg and William S. Laughlin, "Human Adaptability and Achievement," *Culture: Man's Adaptive Dimension, op. cit.,* p. 266, for a critique of such theories.

[41]*See* René Dubos, *So Human an Animal* (New York: Scribner's, 1969), p. 96, for a brief survey of opinions in support of this view.

[42]A. Irving Hallowell, "Self, Society, and Culture," *Culture: Man's Adaptive Dimension, op. cit.,* p. 201.

[43]*Ibid.,* p. 201–202.

[44]*Ibid.,* p. 203.

[45]*Ibid.*, p. 225.

[46]Clifford Geertz, "The Impact of the Concept of Culture on the Concept of Man," *New Views on the Nature of Man, op. cit.*, p. 96.

[47]Clifford Geertz, "The Transition to Humanity," Sol Tax (ed.), *Horizons of Anthropology* (Chicago: Aldine Publishing Co., 1964), p. 46.

[48]Geertz, "The Impact of the Concept of Culture...," *op. cit.*, p. 106

[49]*Ibid.*, p. 116.

[50]*Ibid.*, p. 112.

[51]Clifford Geertz, "Religion as a Cultural System," Michael Banton (ed.), *Anthropological Approaches to the Study of Religion* (New York: Frederick A. Praeger, 1966), p. 13.

[52]Gorden W. Allport, *Becoming* (New Haven: Yale Paperbound, 1967), p. 19.

[53]Martin Buber, "What is Man?," *Between Man and Man*, trans. Ronald Gregor Smith (New York: Macmillan, 1965), p. 123.

[54]*Ibid.*, p. 205.

[55]Columba Ryan, O. P., "The Traditional Concept of Natural Law: An Interpretation," Illtud Evans (ed.), *Light on the Natural Law* (Baltimore: Helicon Press, 1965), p. 23.

[56]Ernest Becker, *The Structure of Evil: An Essay on the Unification of the Science of Man* (New York: George Braziller, 1968), pp. 387–388.

[57]Fromm and Xirau (eds.), *op. cit.*, p. 9.

[58]Marjorie Grene, *Approaches to a Philosophical Biology* (New York: Basic Books, 1968), p. 48.

[59]Erich Kahler, "Culture and Evolution," *Culture: Man's Adaptive Dimension, op. cit.*, p. 14.

[60]Grene, *The Knower and the Known, op. cit.*, p. 217.

[61]Polanyi, *op. cit.*, p. 395.

Human Being

The Boundaries of the Concept*

Lawrence C. Becker

I. Problems of Definition

Uncertainty about our ability to define the biological boundaries of human life is familiar. Currently, the most prominent issue is the definition of death—specifically whether to retain the traditional cardiopulmonary criteria for death or to adopt some version of so-called brain-death criteria. The law in some jurisdictions has already begun to permit physicians to pronounce death on a finding of "irreversible coma." And though it is clear that transplant surgery and the development of life-support technology have given impetus to the change, a number of writers have taken pains to argue that it is perfectly sound, conceptually, to redefine death.

Problems with the definition of the beginning of human life are even more frequently rehearsed. There are advocates of the biological life cycle account, various developmental views, theological ensoulment theories, and "personhood" defintions. The US Supreme Court has recently accepted the view that no conclusive reasons can be found for settling on any one of these rather than another.

The importance of these definitional questions for moral philosophy is obvious. Human beings protect themselves with a thicket of rights they do not grant to other beings, and some of these rights are said to be *human* rights—rights one has simply by virtue of being human. Any conceptual uncertainty about when an entity has become or has ceased to be human is a problem for the ascription of such rights. Further, what might be called threshhold homicides—the killing of entities whose claim to being human is somewhat in doubt—have become increasingly problematic. There are intraspecies threshold questions (abortion, infanticide, some types of euthanasia) and interspecies threshold questions (the killing of other intelligent life forms).

I am concerned, here, with two propositions about the boundaries of human life, each of which has a direct bearing on current controversies and perennial moral problems:

(1) That there is no decisive way to define, in purely biological terms, either the point at which a human life begins, or the point at which it ends.
(2) In any case, if the end points are going to be used as moral divides, they should be defined in terms of morally relevant characteristics, not purely biological ones.

My purpose is to attack both of these propositions by proposing what I take to be decisive biological definitions of the boundaries and by giving reasons for thinking that, for moral theory, such biological definitions are preferable to "morally relevant" ones. The arguments on the latter issue are fairly straightforward and need not be abstracted in this introduction. But the arguments for the boundary definitions are a bit tortuous, so it may be worthwhile to give an overview of them.

The line of argument for the becoming/being boundary may be summarized as follows:

(1) Entry into the class of human beings is a process.
(2) The entry process is at least in part a biological one.
(3) The completion of the biological part of the entry process is a necessary condition for the completion of the entry process *per se*.

(4) The biological part of the entry process is developmental in nature—the development of a set of living cells into a multicellular organism.

(5) The developmental nature of the biological part of the entry process is best understood by way of an analogy with metamorphosis—that is, as the genesis, from the relatively undifferentiated mass of the fertilized ovum, of the fundamental morphology and histologically differentiated organs the organism is genetically programmed to develop.

(6) The completion of what I shall call the metamorphic phase of generative development is a necessary condition of the completion of the entry process—that is, the becoming/being boundary cannot be put any earlier than this.

(7) There are no good reasons for putting the boundary any later than this.

(8) Therefore, the becoming/being boundary lies at the completion of the metamorphic phase of generative development.

The line of argument for the being/has-been boundary is parallel:

(1) Exit from the class of human beings is a process.

(2) The exit process is at least in part a biological one.

(3) The completion of the biological part of the exit process is a necessary condition for the completion of the exit process *per se*.

(4) The biological part of the exit process is disintegrative in nature.

(5) The distintegrative nature of the biological part of the exit process is best construed as the functional disintegration of the organism as such—and not as the physical disintegration of its parts.

(6) The completion of the disintegration of the organism as such is a necessary condition for the completion of the exit process—that is, the being/has-been boundary cannot be put any earlier than this.

(7) There are no good reasons for putting the boundary any later than this.

(8) Therefore, the being/has-been boundary lies at the completion of the disintegration of the human being considered as a biological organism.

Without further ado, then, I shall turn to the arguments for the becoming/being boundary.

II. The Becoming/Being Boundary

A caterpillar is not a butterfly. That is, the insect of which the caterpillar is the larval stage is not, at the larval stage, a butterfly — though one might, as indeed biologists do, speak of butterflies as "adult butterflies" in order to emphasize the fact that both caterpillars and butterflies are stages in the development of the same insect. Nonetheless we do not confuse insects which *are* butterflies with insects of the same species which *are* catepillars. The latter are *becoming* butterflies no doubt, but they are not butterflies yet.

When can we say that the insect *is* a butterfly as opposed to a caterpillar (or rather, a pupa)? Surely we can say this only when the process of metamorphosis is complete—that is, when the relatively undifferentiated mass left by the disintegration of the caterpillar's tissues has metamorphosed into the pattern of differentiation we call a butterfly.

Human fetal development is a process analogous to metamorphosis, and just as it makes good sense to speak of butterfly eggs, larvae, and pupae as distinct from the butterflies they become (to say that they are *not* butterflies) so too it makes sense to say that human eggs, embryos, and fetuses are distinct from the humans they become— that they are not human *beings,* only human becomings.

When can we say that the fetus is a human being rather than a human becoming? Surely only when its metamorphic-like process is complete—that is, when the relatively undifferentiated mass of the fertilized human ovum has developed into the pattern of differentiation characteristic of the organism it is genetically programmed to become.

That is the core of what I have to say about the becoming/being boundary. But it will require considerable elaboration and defense, and it may help to note, to begin with, that the definitional problem here is to clearly describe a concept of "being"—a static, or at any rate reasonably stable, "completed" condition. This is not to say, of course, that human beings are themselves static or unchanging. It is merely to indicate that we are looking for the boundaries that define membership in the class of living humans—and which distinguish that class from the class of entities that might, but have not yet, become humans, as well as from the class of entities that have been, but are no longer, humans. In the case of the becoming/being boundary, then, we are looking for a point at which the entity is in some very fundamental sense "completed" as a member of the species. I shall argue in what follows that this point is reached when the organism (assumed, of course, to be living) has assumed its basic morphology, and when its inventory of histologically differentiated organs is complete. (It may be worth pointing out one subtlety here. I will argue that the process is complete for a given organism when that organism's inventory of organs is complete—not when some standard list of human organs is filled. This is done to account for mutants.)

The rationale for this point as the becoming/being boundary begins with the straightforward observation that entry into the class of living human beings is a process. The claim that "entry is a process" means no more than that humans come into being *by way* of a process. This process is, at least in part, a biological one—involving at a minimum the production of an ovum in a suitable environment for parthenogenesis or cloning, and typically the production of both ovum and sperm, together with the processes necessary for their union. Whatever else we may want to say about this process of entry, we have to concede, surely, that the completion of its biological aspects is a necessary condition of its completion *per se*. (Whether it is also a sufficient condition will be discussed later.) Thus it is clear that the becoming/being boundary cannot be put at a point prior to the biological completion of the process of entry.

The starting point of the process is not in dispute here, though to put it at conception would beg an important question. So assume that the process starts well before conception—say, with the production of the particular ovum which is to be fertilized (or perhaps "activated" in the case of cloning). The question to be answered, then, is: At what point do we have adequate reasons for saying that the process is biologically complete?

A standard answer is derived from the concept of a "life cycle." The argument is that the life cycle of a human being begins at conception—just as the life cycle of a butterfly begins with a fertilized egg, progresses through the larval and pupal stages, and culminates in the development of what is popularly described as a butterfly.

The trouble with the caterpillar/butterfly analogy as proposed above —according to the life cycle argument—is that it misleads one into thinking that entry into the *species* coincides with the end of metamorphosis. Quite the contrary: egg, larva, pupa, and butterfly are all stages in the development of *one* entity of *one* species; just as conceptus, embryo, fetus, neonate, infant, child, adolescent, and adult are all stages in the development of an entity of the species *Homo sapiens*. "Being" a human thus begins at conception—at the beginning of the life cycle.

This is a rhetorically persuasive argument, but it contains both logical and empirical errors. The fundamental logical error can be seen most clearly by first considering the obviously fallacious syllogism (all too frequently taken seriously):

This conceptus is a being (i.e., is an entity and is alive).
It is certainly human (i.e., is of no other species).
Therefore, it is a human being.

The fallacy here is equivocation on the word "human." As used in the premise it is an adjective—and as such applies not only to the conceptus but to any living part of a member of the species: human blood, human sperm. But as used in the conclusion, "human" functions as a noun meaning "member of the species *Homo sapiens*." A counterexample will suffice to make the point.

This sperm cell is a being (i.e., is an entity and alive).
It is certainly human (i.e., is of no other species).
Therefore, it is a human being.

The fallacy in the life cycle argument is not quite as blatant, but it is similar. From the premise that fertilization of the ovum produces a unique living entity which is a product of the species, it does not follow that that entity is a *member* of the species. It is just as possible to conclude that the entity produced by fertilization is one which will *become* a member of the species.

The empirical error in the life cycle definition utterly destroys its plausibility as an account of the becoming/being boundary. Monozygotic twinning can occur any time from the two-cell stage to about the 14th day after conception. And it is thought that most such twinning is not genetically determined.[1] What this means is that one cannot say at conception, even given complete knowledge of the genetic makeup of the conceptus, how many humans will develop from it. It surely will not do, therefore, to say that the process of becoming a human being ends at conception.[2]

But if not at conception, then when? Shall we say that the becoming/being boundary lies at the point where the number of embryos is irrevocably determined? Shall we say that the life cycle of a human being begins at that point? I think not—this time for purely conceptual reasons.

It has already been noted that there is no logical necessity in the inference from the premise that a unique, living, and human entity exists to the conclusion that that entity is a human being—i.e., a member of the species as opposed to an entity in the process of becoming one. So we are certainly *not forced* to put the boundary at the end of the twinning possibility. Indeed, I suggest that when we reflect on the nature of human development, the only point for the becoming/being boundary which makes conceptual sense is at the end of what might be called its metamorphic phase. Some detail will be helpful at this point.

"Biological development" is a very broadly defined term. One writer says: "Development may be defined as the action of genes in: (1) creating a new organism from some part of a parent organism, (2) maintaining or increasing the size of a fully formed mature organism, and (3) repairing accidental defects or losses in an organism...."[3] It is clear that the first category above is the sort of development of concern to us here. Let us call it (human) *generative* development, to distinguish it from the other sorts that are typically referred to as continuous development and regenerative development, respectively.

Human generative development involves four sorts of processes: (a) cell proliferation, in which the number of cells increases; (b) growth, in which there is an increase in the mass of the developing organism; (c) morphogenesis, in which progressive changes in form take place; and (d) histogenesis, in which cells specialize into tissues. Morphogenesis and histogenesis are often lumped together under the title *differentiation*.

Differentiation, cell proliferation, and growth are all involved in continuous and regenerative development as well as in generative development, of course. The continuous production of red blood cells throughout life is an example of histogenesis. Obviously, to speak of the sort of completion indicated by the becoming/being boundary is not to speak of the completion of such processes of maintenance and regeneration—however similar they are in kind to the processes of generative development. It is rather (at least in part) to speak of the completion of the process of the generation of a new (human) organism.

But what is the nature of this process, and when is it complete? While the distinction between the generation, maintenance, and re-generation of an organism is reasonably clear at the most abstract level, how is one to give it application in the case of human development? It is here that the analogy to metamorphosis is helpful.

In common biological usage, "metamorphosis" doubtless refers to the sort of transformations undergone by the developing butterfly—where there is first the generation of a free-living larval body distinctly different from the adult, then the de-differentiation of the tissues of that body and the subsequent generation of the adult. But biologists who have addressed themselves to examining the nature of this process characterize it in a way which, without strain, fits human fetal development. Embryonic and metamorphic development are often spoken of conjointly.[4] In fact, as one writer classifies types of metamorphosis, a distinct larval body is not required at all—and thus human generative development sits comfortably as a *type* of metamorphosis.[5]

Now it is clear that generative development has both what might be called *fundamental* aspects and aspects that are essentially refinements of or maturation of the basic structures of the organism. The neonate has a skeletal system of about 270 bones. "Fusion of some of these in infancy reduces this number slightly, but from then until puberty there is a steady increase...at puberty there are 350 separate bony masses, and this number is increased still further during adolescence. Thereafter, fusions again bring about a reduction to the final quota of 206..."[6] Similarly with gametogenesis. Oögonia in the female and spermatogonia in the male are present before birth, but their maturation into full fledged ova (i.e., oötids) and spermatozoa only comes about at puberty. In the case of the lungs, the alveolar ducts are present in the fetus, but only after birth do the alveoli proper develop and they continue to proliferate well into the eighth year of childhood.[7] Examples of such refinement and maturation of structures—undeniably a part of generative development—could be multiplied.

But the original analogy to metamorphosis is instructive here. Just as it is not the size of the entity, or whether its cells are proliferating, which is at stake in our judgment that the pupa has become a butterfly, so too we are not concerned with various refinements, adaptations to environment outside the cocoon, and maturation which might take place in the butterfly's basic structure. Metamorphosis—*at least in the sense relevant to drawing the line between pupa and butterfly*—is complete once these basic structures are complete. Similarly for humans. Generative development in the form of refinements, adaptations, and maturation of the basic structures are not of concern in drawing the becoming/being boundary.

But what counts as the "basic structure" and when is its generation complete? This is probably a more difficult question empirically than it is conceptually. Conceptually, the answer is not hard to find. The metamorphic phase of generative development (i.e., the "fundamental" differentiation) is complete when (1) the organism has assumed its basic gross anatomical form, normal or not (by which I mean its basic skeletal structure, musculature, arrangement of organ masses, and distribution of tissues); (2) the organism's inventory (normal or not) of histologically differentiated organs is complete.

This is not, notice, a functional criterion so much as an anatomical one. That the developing embryo is "alive"—i.e., functioning as a biological organism—is assumed. The question is when, in the course of its development, we may say that its fundamental or metamorphic generative development is at an end.

It seems indisputable that the end cannot be put any earlier than the point described above. After all, if anything is basic to human generative development (beyond conception), it is the shaping of the formless mass of cells into the shape and general arrangement of parts which the continuous and regenerative processes of development will maintain. And an "organ" which is not histologically differentiated is no organ at all. I do not think anyone would want to hold that the generation of organs was not a part of basic generative development. So the boundary can surely not be put any earlier than the point I have described. And to put the point later than that—to require, for example, that the differentiation of the ciliary muscles of the eye be complete—stretches the notion of fundamental or basic structures beyond reasonable bounds. I do not mean to claim that the distinction between basic and nonbasic generative development is conceptually crisp—such that, given any example of generative development, it could be unarguably classified in one and only one of the categories. I merely hold—and will argue below—that the distinction is clear enough for the use we need to make of it.

The empirical question, however, may be more difficult—or at least it seems so to a nonbiologist reading the standard sources. The completion of gross anatomical form is not much of a problem. That is virtually complete by the end of the third lunar month of gestation —so much so that aborted fetuses of that age can be used in place of cadavers to teach anatomy to medical students. "...[W]ith the aid of a simple magnifier, every gross anatomical detail can be seen."[8] Further changes in morphology (e.g., as late as those occurring during puberty) are either the regional growth of existing structures, or clearly in the category of refinements, adaptation, and maturation of those structures.

The histogenesis of organs is a more difficult matter. It is clear that very few organs are histologically differentiated at the end of the third lunar month. Indeed, development of the alveolar ducts and the formation of elastic tissue in the lungs occurs well into the sixth lunar month.[9] Parts of the digestive system (e.g., esophageal glands) are defined even later.[10] The timetable for these later developments is not terribly precise, and no doubt can never be, because of individual variations. But it seems true to say that the end of what I am calling the metamorphic phase of generative development can be put no earlier than the middle of the sixth lunar month of gestation and need not be put any later than the middle of the final month—generative development thereafter clearly falling into the refinement, adaptation, and maturation category. (The various skeletal rearrangements, the myelization of neural tissue, the proliferation of alveoli, and gametogenesis are all clearly in the nonbasic category.)

Suppose, then, just for the argument, that we say that the metamorphic phase of generative development is complete at the beginning of the eighth month. Are we then in a position to defend the claim that such a fetus is a human *being* as opposed to a human becoming? Reasons have been given for the contention that the becoming/being boundary cannot be put any earlier than this—that is, that the completion of generative metamorphosis is a necessary condition for entry into the class of human beings. But is it also a sufficient condition? Are there good reasons for thinking that the completion of the metamorphic phase of generative development is enough to count as crossing the becoming/being boundary? I think there are good reasons—conclusive ones in fact—but they are of a negative sort. That is, I think the reasons consist in there being no good reasons for requiring anything further by way of a condition. The clearest way to show this is by dealing with some obvious objections to the metamorphic definition as here proposed.

III. Objections to the Metamorphic Definition

Imprecision

It may be argued that the obvious imprecision in the timetable of metamorphosis is intolerable, as one cannot know in advance —at various points in the last few months of gestation—whether, for example, a particular abortion will be homicide or not.

The reply to this objection may be brief. We are faced with many such uncertainties in both the law and morality. Often we have to deal with a process and need to know precisely when it was "complete," but find the difficulties nearly insuperable. Consider the notorious difficulties of distinguishing an attempted crime (an indictable offense) from the mere preparation to attempt it (which is not an indictable offense).[11] Such problems cannot be solved, they can only be handled. And in the case of the definition of "human being,"*if it is to figure in the administration of a stringent prohibition of homicide*, it seems reasonable to adopt an empirically conservative presumption. If basic, generative differentiation has ever been known to be (or can reasonably be thought to have been) complete by the end of the first week of month seven, then one might invoke the presumption of homicide for the destruction of any fetus reasonably believed to be in or beyond the seventh month of gestation. Or one might want to adopt a series of increasingly strong standards of care, beginning at the latest point at which the process can be guaranteed to be incomplete.[12] In any case this is a practical problem of a sort endemic to law and morality, and it is safe to say that the leading alternative candidates for the becoming/being distinction (i.e., conception, viability, and the development of personality) are also subject to it. It cannot, therefore, constitute any special objection to the metamorphic definition.

Mutation and Arrested Development

A critic may want to know more, however, about how the definition handles cases of mutation and arrested development. What about the fetus which develops no limbs, or only one kidney, or a heart with only three valves?

Here it helps to remember that the metamorphic definition—beyond requiring genetic material from the species—is phrased in terms of the development of each individual. Whether that individual has a genetic anomaly which causes a mutation in form or organ inventory, or whether environmental factors put a premature end to development is irrelevant. If the fetus (mutant or not) dies or is killed before the completion of the metamorphic phase of *its* generative development, then what has died or been killed is a human becoming. If the fetus survives, and the process of differentiation is complete, yet the fetus is not normally formed, then what lives is a non-normally formed human being. If the fetus is born prior to the completion of the process, but given the proper environment, can survive while the process continues to completion, then what has been born is a human becoming. It should be noted that none of this implies, by itself, the existence or nonexistence of specific duties toward such fetuses. The morality of the treatment of fetuses of various sorts and in various stages of development is a matter for further argument. It is no objection to the metamorphic definition to show that it does not settle such matters.

Alternatives to the Metamorphic Definition

The first two objections aside, there may be some remaining feeling that the choice of the metamorphic definition is as arbitrary as several other alternatives. Even if conception and the terminus of the twinning possibility have been ruled out, why not choose the concept of viability—on the grounds, perhaps, that a human being is not a biological parasite, but that the fetus is just that until the point of viability? Or why not choose quickening or live birth or the development of personality? To relieve this dissatisfaction, it will be necessary to say a few words about some of the other standard candidates for the becoming/being boundary.

The viability alternative is unsatisfactory. It confuses a criterion with a definition. Viability is not a *definition* of "human being." One can, after all, have a nonviable (but temporarily alive) human being. Viability is rather, in fact, a rough *criterion* for the completion of the process of metamorphosis. Viability (outside the mother's body and outside mechanical facsimilies of it) coincides— roughly—with the end of basic histogenesis as here described.

Other alternatives to the metamorphic definition have even less plausibility. "Quickening" has nothing to recommend it even initially—unless it is confused with viability. The point of "live birth" is flatly arbitrary, bearing as it does no necessary relation to properties of the fetus. It has some advantages as a legal device for fixing age, but beyond that, has nothing to recommend it.[13] And the development of "personhood," as a definition of human being, only has interest if one is singlemindedly trying to build up a definition which will yield "rights to life"—where such rights are understood to arise only from the claims one agent may make on another. It taxes the concept of membership in the species too far to say that a 14-year-old, so catastrophically deficient as to warrant the claim that he or she is not a "person," is not a member of the species.

The Moral Emptiness of the Definition

But then, it may be urged, one has abandoned any attempt to make the becoming/being boundary a moral divide. One can understand how quickening might be held to have characteristics relevant to a moral boundary—for it has a psychological impact on the pregnant woman and others. Similarly with live birth and the development of personality. But "the end of the metamorphic phase of generative development" does not seem to capture any morally significant distinction. And the resistance to adopting a morally empty definition, given our actual use of rough and ready becoming/being boundaries as moral divides is strong. As Tooley and others have argued, if the legitimacy of moral prohibitions and permissions (say, with regard

to killing) are going to rest on whether or not the victim has crossed
the becoming/being boundary, then the drawing of that boundary
must be done in terms of characteristics relevant to the moral justification of those prohibitions and permissions.[14] This is an important
line of argument, so I want it to be clear why I reject its applicability
here.

In the first place, I think we may plausibly reject quickening and
live birth as candidates for the sort of moral divide at stake here. We
are, after all, talking about *duties* not to kill, and the sort of psychological pulls created by these two events (aside from the fact that not
every parent will feel them) are just not the sort of grounds advocates of a morally relevant definition are interested in. They are
interested in justifying a right *in the victim* not to be killed—a so-
called right to life.

Now if one tries to derive the moral rules concerning homicide
from special rights to life possessed by the victims and wants to
show that those rights to life are derived from some characteristics
which define the victims as human beings, then the metamorphic
definition is indeed beside the point. So, I believe, are all other non-
theological definitions except personhood. The question really is,
then, why not adopt personhood as the becoming/being boundary?
Even if it leads to unpleasant conclusions such as a failure to rule out
infanticide, at least it marks a moral divide of major proportions.
Persons—more exactly, self-conscious subjects of experience—can
value themselves. And in just the same sense in which my values for
my act *A* support the rationality of that act *A,* so too another person's
values *against* (my act) *A* support the rationality of (my act) *non-A.*
Thus there is one clear sense in which persons can make claims on us
which nonpersons cannot make. And since the making of such
claims has an obvious application to the question of homicide, it is
tempting to try to base one's account of the morality of homicide on
such claims.

But I think it is not usually recognized just how unsatisfactory this
whole approach is. For one thing, though an obviously sound basis
for moral argument, it is a very slender reed on which to hang the
whole analysis of homicide. To suppose that all our duties not to kill
come from the victim's *rights* (to life, liberty, or whatever), and that
those rights are grounded in the victim's ability (and title) to claim
certain acts and forbearances from others, is to put oneself in a very

awkward position theoretically—not to say morally. What account is one to give, then, of a parent's duties to his or her infant offspring? What account is one to give of our duties not to kill the sleeping? Or temporarily comatose? Or our duties to resuscitate those who have drowned? One is forced, on this account, either to deny the existence of such duties or to construct an account of how such duties can arise from counterfactual conditions (i.e., if *B* were awake, or at the age of reason, or alive again, he would lay claim on me for *X*.).

Surely either of these positions is implausible. The counterfactual account is an awkward contrivance in many of these cases. But beyond that, both alternatives ignore some obviously sound lines of moral argument that derive duties from considerations which begin with the *agent,* rather than with the one acted upon. A duty not to kill (or a duty to rescue) may be justified by reference to the consequences for the agent or society. It may be justified as an entailment of the agent's role (parent, doctor, friend). Or it may be justified as a requirement of those patterns of life or character traits of which we can justifiably approve, morally. None of these justifications makes essential reference to the victim's ability and title to lay claim to the duty.

Now I am not suggesting that such agent-based approaches can, by themselves, be any more adequate than the victim-based approach. A general account of the morality of killing which ignored the victim's claims on the agent would be indefensibly incomplete. But so is an account which ignores the agent-based approaches. And once the need for both sorts of approaches is recognized, the attempt to rig a definition of "human being" along the lines suggested by *any* single line of argument (whether victim-based or not) seems arbitrary in the extreme. There can, for example, be no *a priori* guarantee that the range of entities protected by duties generated from agent-based approaches will coincide exactly with those protected by duties generated from victim-based approaches. So at the least it is certainly an invitation to question-begging to force the terms "human being" and "homicide" into the area circumscribed by the victim-based approach.

Further, of course, the question of homicide not only involves threshold problems (i.e., whether the victim is a human being or not). It also involves giving a rationale for retaining or rejecting all the intricacies of homicide law—the grading of various sorts of homicide, the exculpatory claims we recognize, and the category of justifiable homicide. Any "right to life"which could conceivably be encapsulated in a definition of "human being" would prove an infertile ground indeed for these matters. Consider: appeals to personhood are of no avail in explaining the distinctions we draw between deaths produced by tortious negligence, criminal negligence, and premediated acts of murder. Human victims of each have, one assumes, an equal "right to life," and surely, *with respect only to that right,* no less a claim on their fellows for reasonable care as for non-malicious conduct.

It is, of course, possible to build up an account of the details of homicide law by reference to other principles, using the "right-to-life" notion merely as a starting point. But then one must acknowledge, surely, that the "right to life" is itself very nearly vacuous, morally. It functions as nothing more than a general presumption against a certain restricted class of morally problematic killings, and even then it is not relevant to deciding many of the questions we need answered about homicide. This, together with the difficulties of even explicating any morally relevant definition of the "point of entry into humanness" shows, I think, that the objection of vacuousness against the metamorphic definition is without much force.

Indeed, I reiterate that the primacy of the right-to-life line of argument is a snare. A much more straightforward, and thus conceptually clear, approach is simply to ask what presumptions against the taking of life there are, and why, and under what conditions those presumptions may be overcome. A consideration of right-claims made by one agent on another will be a part of this approach, but it is clear that the approach will not be limited to such considerations. Presumptions with regard to the taking of all lives (vegetable, animal, human, potential, or actual) will be confronted directly—not through a mystifying (and doubtless largely self-serving) thicket of special rights definitionally borne by human beings.

This approach to the morality of homicide seems to me to offer more hope of productive, reasoned discussion than do the usual arguments. It will not be easy to specify the grounds for or against a strong presumption concerning the homicide of the fetus of eight months as opposed to a weaker presumption, or none at all, against feticide prior to eight months. But at least the need for argument and the general range of relevant considerations will not be obscure.

One may want to begin with a consideration of the prohibition of homicide in the case of healthy adult victims. One would ask for the justification of the prohibition and for the justification of the various exculpatory claims we allow (or ought to allow). One would then work out to threshold questions, such as abortion and euthanasia, in stages, asking the same questions for each stage. Such a process would be uncomfortable, because it would call into question one of our most central and deeply felt moral principles. But unless wisdom profits from evasion, this is exactly what needs to be done. The definition of the becoming/being boundary bears no *a priori* relevance to this sort of investigation. And if there is a cogent biological definiton of the boundary—as I have argued there is—there is no point in resisting it for the purposes of moral theory.

IV. The Being/Has-Been Boundary

I said at the outset that the definition of "human being" had to separate not only "being" from "becoming' but also "being" from "has been." I want to conclude by deploying an argument to do this—both to complete the promise and to underline my point about the proper approach to the morality of homicide. Given the fervor with which the definition of death is being discussed currently, the brevity of the argument to follow may be perceived as a fault. But I believe that, unlike the becoming/being distinction, the definition of death presents no serious conceptual problems. There are serious empirical problems associated with the clinical determination of when death occurs, and serious moral problems concerning the treatment of the dying and the dead, but those are separate matters. I shall comment on their relation to the definition of death as the argument proceeds.

On the view proposed here a human being is a biological organism, complete as a living "being" of the species when the metamorphic phase of generative development is complete. Death for such an organism is the same as for any other complex organism. It is a process. The process is, at least in part, a biological one. The completion of the biological part of the process is a necessary condition for its completion *per se*. This much I take as not needing argument.

I further take it that we may plausibly regard organic death as the completion of the biological part of the "exit process." This intro-

duces an apparent asymmetry into the account, for the becoming/
being boundary was drawn in terms of structure, not function. But it
should be remembered that the organic life of the developing entity
was presupposed as a necessary condition of "human-beinghood." It
simply was shown not to be a sufficient condition. But just as organic life precedes the generation of the structures necessary for
entry into human-beinghood, so too death precedes the physical disintegration of (most of) those structures. Since life is a necessary
condition for biological entry into human-beinghood, its removal
(death) is sufficient for marking the completion of the biological part
of exit from human-beinghood. The exit process, then, in its biological aspects, is to be construed as a loss of function, not structure.

The being/has-been boundary can thus not be put any earlier than
the biological death of the organism. And I shall assume that human
beings are mortal in such a way that there is no question but that biological death is a *sufficient* condition for marking the being/has-been
boundary. I assume, in particular, that consciousness does not persist beyond organic death.

The interesting question is, of course, What counts as the death of
a human being considered as a biological organism? Clearly, parts
of an organism may die without bringing about the death of the
organism as such. Organisms may lose parts (limbs or organs) and
continue to function organically. They are not "partially dead" for
that reason. They are simply organisms of a certain type without
certain parts. Further, organisms may lose functions necessary to
their survival. If these functions are provided mechanically, and
thus the organism survives as an organism, it is not dead, it is simply
an organism kept alive mechanically.

The biological death of a human organism may be quite straightforwardly described: a human organism is dead when, for whatever
reason, the system of those reciprocally dependent processes which
assimilate oxygen, metabolize food, eliminate wastes, and keep the
organism in relative homeostasis are arrested in a way which the
organism itself cannot reverse. It is the confluence of these and only
these conditions which could possibly define organic death, given
the nature of human organic function. Loss of consciousness is not
death any more than is the loss of a limb. The human organism may
continue to function as an organic system. Further, though the loss
of one vital function (say loss of the capacity to eliminate wastes)

may inevitably *bring about death,* it does not constitute death by itself. Nor would we even say that an arrest of *all* the vital functions, in such a way that the organism *itself* could "restart" them, was death. (Consider the legal fate of one who maliciously intervened to prevent the "restart." Surely we would regard such a person as a murderer, and we would not be speaking metaphorically. On the other hand, when an organism has failed in such a way that it cannot restart its organic processes, *but could be resuscitated by someone else,* what would be the legal fate of one who maliciously refused to resuscitate? Surely not an indictment for murder.)

Now it may be objected that requiring the confluent cessation of *all* the organic functions mentioned is too strong. First, they usually do not cease simultaneously, and second, it would be somewhat strange to withhold the judgment of death from an organism whose sole remaining organic function was some waning remnant of the digestive process, such as the action of enzymes in the intestines. True. But the definition proposed here does not entail such a result. Death is defined as the conjoint (not necessarily simultaneous) cessation of *the system* of those reciprocally dependent processes which assimilate oxygen, etc. Some of these processes involve the production of biochemical agents (e.g., enzymes) which, as to their continued existence and operation, are then relatively independent of the processes that produced them. But the continued action of such agents, in the absence of the process which produced them, cannot properly be considered the continuance of the process. It is rather the action of isolated remnants of a process which has itself disintegrated. There are many such events which continue as artifacts of vital processes after death. A cell may live, though the organ of which it is a part is dead (i.e., no longer functions as an integrated subsystem of an organism). An organ or tissue may remain functional for days after the death of the organism as a whole (as with the cornea or blood removed from the body or skin kept protected from bacteria). None of these events embarrass the definition of death given here.

It should be emphasized, however, that this definition of death is to be sharply distinguished from the notion of a clinical criterion for the death of a given individual. When we may correctly say that an organism has ceased to function as an organism in the requisite sense is an empirical problem of considerable delicacy. Fortunately for moral purposes, the functional disintegration of the human organism (if not mechanically assisted) is marked by reasonably unambiguous clinical signs whose "appearance" (e.g., the registering of cardiopulmonary failure) takes a relatively short duration. So no one exercising reasonable care is likely to have to rush the determination. (Certain emergency situations are, of course, exceptions.)

Where mechanical assistance is provided to maintain organic function, the implications of the definition of the human being/has-been boundary are fairly clear. One whose heart no longer functions and who is kept alive by machine is just that—a human being whose heart does not function. One who, after a massive accident, has a flat electroencephalogram and no spontaneous respiration, heart activity, or kidney function, and whose organs are bypassed or kept functioning by heroic medicine is just that. The definition makes no reference to the "higher" functions characteristic of humans or to how organic function is maintained. (After all, in the ninth month of gestation, not very many "higher" functions are going on, and the fetus functions as an organism partly by virtue of assistance provided by the mother's body.)

This is, surely, not only a common-sense view, but one which faces the moral problems raised by heroic medicine and euthanasia directly. The moral question here is not whether the permanently comatose are "really human." The question is, under what circumstances ought one to use heroic measures on humans who would otherwise die, and once in use, under what circumstances may they be withdrawn? Similarly for questions of "positive" euthanasia. Much clarity is lost, I think, by organizing inquiries into these matters in terms of a definition of "human being" which settles the issues. Such definitions merely push the important questions back one notch (or worse, allow people to evade them), and inevitably seem *ad hoc* in nature.

Locutions such as "brain death" are thus misleading when construed as definitions of death. Brain death is not a definition of death, nor even a criterion of death. It is merely a criterion for deciding when coma is irreversible.[15] The moral question, accurately put, is, What should be done with human beings who are in irreversible coma?

There is considerable pressure to resist this conclusion and to allow physicians to pronounce death upon a finding of irreversible coma.[16] The motives behind the move are not hard to discern. Beyond a point that can be specified empirically with some accuracy, hope for bringing the patient back to any form of consciousness—no matter how rudimentary—is simply gone. The brain literally liquifies. And even with the most sophisticated mechanical aids, the other vital organs begin a slow but certain course of degeneration. Leaving aside the desire of some for organs suitable for transplantion, it is an enormously expensive and futile effort to keep such hopelessly comatose patients alive.[17] To be able to pronounce them dead would be a great convenience. It would eliminate any legal hazards involved in "pulling the plug" (for if such patients are regarded as living, turning off the respirators or other devices already in use amounts to active, rather than passive, euthanasia—to killing rather than to letting die). There are, in most cases, no legal obligations to begin such treatment (no legal duty to rescue); but there are often moral obligations, because it is often not clear before the efforts are made whether or not the patient is in irreversible coma. The irony is that once treatment is begun, there is often a legal obligation to continue, although there may be no moral obligation to do so.

Rigging the definition of death to solve this problem, while tempting, is an avoidance of the real issue. The real issue is whether and, if so, when it is moral to give up trying to prolong the patient's life. Putting this question in terms of euthanasia or even "letting people die" is a misleading sensationalization of the issue. Euthanasia is certainly an important moral question in its own right, but the typical medical situations—at least the ones in which the temptation to bring in the definition of "human being" arises—are those in which efforts to prolong life are underway, and the question is whether it makes sense to go on with them. "Giving up" is not always irrational or immoral—and certainly not always illegal.[18] It seems best to face this problem directly—by defining when it is permissible to give up life-saving efforts—and not to evade the problem by introducing an *ad hoc* definition of death.

The being/has-been boundary thus should not be, by itself, a moral divide any more than the becoming/being boundary is. People live, but sometimes in such hopeless conditions that one may morally and legally give up trying to save them. People die, but sometimes can be revived. Their death does not in itself relieve us of moral obligations toward them.[19] The reversibility of death is more likely the moral divide. But even irreversible death does not (under our ordinary convictions) mean we can do just as we please with what remains—e.g., the estate, the body. The morality of dealing with the dead, whether reversibly dead or not, is a matter for further argument. It is not settled by this definition of death.

Bizarre questions may be raised, of course. Is a human brain separated from its body and kept "functioning" a human being? (Assuming that the removal of limbs or eyes or heart and lungs would still "leave" a human being.) I admit to being at a loss for a reply to such cases, let alone an answer.

But the inability of a definition to settle bizarre cases need not be considered an overwhelming defect. Bizarre cases can often be settled only by equally bizarre definitions. The definitions proposed here—for both the becoming/being and the being/has-been boundaries—make good sense conceptually, are sufficiently clear for moral purposes, and direct our attention to the moral issues surrounding homicide in a productively direct way. That much, it seems to me, is enough to expect from definitions.

To summarize the conclusions, then, from the somewhat tortuous path just trod, I have argued that:

(1) There are rationally preferred choices for both the becoming/ being and being/has-been boundaries, drawn in purely biological terms. The former boundary lies at the completion of the metamorphic phase of generative development; the latter at the functional disintegration of the human being considered as a biological organism.

(2) Neither of these boundaries is, by itself, a moral divide.

(3) Each of the boundaries is precise enough for use as a moral divide if further argument esablishes the legitimacy of it.

(4) Such further argument cannot reasonably be only of the "victim's right-to-life" variety.

(5) As it turns out, on the abortion question, the US Supreme Court's advocacy of graduated stages of state interest fits the becoming/being boundary reasonably well.

(6) "Brain death" is neither a definition of, nor a criterion for, the being/has-been boundary.

Doubtless other conclusions are implicit in the arguments. For the moment I content myself with these.

Notes and References

*Earlier versions of this paper were read to the Conference on Moral Problems in Medicine, sponsored by the Council for Philosophical Studies, and to the Philosophy Club of the University of Virginia. Thanks are due to members of both groups for helpful comments, but my particular gratitude extends to Jean W. Hitzeman, James E. Kennedy, Marvin Kohl, and George M. Brockway, who saved me from some serious errors.

[1]*See* M. G. Bulmer, *The Biology of Twinning in Man* (Oxford, 1970), and relevant passages from Max Levitan and Ashley Montagu, *Textbook of Human Genetics* (New York, 1971). The importance of the issue of twinning was brought to my attention by James M. Humber's paper, "The Immorality of Abortion," presented at the Eastern Division Meetings of the American Philosophical Association in December 1973. It should be noted, however, that at least in terms of the arguments of that paper, Mr. Humber would apparently not agree that this problem is a significant one for the conception definition.

[2]One can, of course, assert the contrary by saying that once conception occurs, nonbiological souls come into being, and that the number of souls thus brought into being determines (or corresponds to) the number of fetuses which will develop. One would then have to go on to identify the existence of the souls with the existence of human beings. But one can assert the contrary of any proposition whatsoever in this way—assuming the assertion is not self-contradictory. That is, one can merely invent an alternative. The question is, what *reasons* can be offered to support the truth of such an alternative claim? I shall assume here—and it is surely not an arbitrary assumption—that no philosophically defensible reasons can be found to support the "nonbiological soul" alternative in this context.

[3]Nelson T. Spratt Jr., *Developmental Biology* (Belmont, CA, 1971), p. 5

[4]*Ibid.*, p. 17; and the following: "Metamorphosis is a widespread developmental phenomenon which is usually associated with a dramatic change in habitat and consequent way of life....Primarily it consists of the differential destruction of certain tissues, accompanied by an increase in growth and differentiation of other tissues. The phenomenon of regional growth and differentiation associated with local cell death in developing limbs comes into this category." N. J. Berrill, *Developmental Biology* (New York, 1971), pp. 423–424.

[5]Spratt, *Developmental Biology,* pp. 283-284, quoting Weiss, *The Science of Zoölogy* (New York, 1966).

[6]L. B. Arey, *Developmental Anatomy,* 7th ed. (Philadelphia, 1965), p. 405.

[7]*See* J. B. Thomas, *Introduction to Human Embryology* (Philadelphia, 1968), p. 297.

[8]Hans Elias, *Basic Human Anatomy as Seen in the Fetus* (St. Louis, 1971), p. vii.

[9]"Primary ossification centers of the pharyngeal arches appear...the circular ciliary muscles of the eye are differentiation...The lumina of parotid and sublingual glands are established...primordia of Peyer's patches appear in the ileum....Development of the alveolar ducts including the formation of elastic tis-

sue is prominent....The hyaloid artery of the eye begins to degenerate." J. B. Thomas, *Human Embryology,* pp. 280–281.

[10]*Ibid.*, p.297.

[11]*See,* for example, a standard handbook on the substantive criminal law: Wayne R. LaFave and Austin W. Scott Jr., *Handbook on Criminal Law* (St. Paul, 1972), pp. 431-438. For a review of some of the cases and comment on the philosophical aspects of the problem, see my article, "Criminal Attempt and the Theory of the Law of Crimes," *Philosophy & Public Affairs* **3,** no. 3 (Spring 1974), 262–294.

[12]The Supreme Court has done something similar in its recent abortion decision. *See* Roe v. Wade 410 US 113, 41 LW 4213 (1973) at 4214.

[13]That is, there is nothing to recommend it as a becoming/being boundary. As a moral distinction based on the fact that the neonate immediately begins "to play an explicit role within the social structure of the family and society" there may be more to say for it. *See* H. Tristam Engelhardt, Jr., "The Ontology of Abortion," *Ethics* **84** (1974): 217–234, especially pp. 230–232.

[14]*See* Michael Tooley, "Abortion and Infanticide," *Philosophy & Public Affairs* **2,** no.1 (Fall 1972), 37–65.

[15]The Harvard Medical School panel charged with defining "brain death" conflates these questions misleadingly. *See* their report in Henry K. Beecher, *Research and the Individual* (Boston, 1970), pp. 311–319.

[16]As recommended by the Harvard panel, *ibid.,* and as is beginning to get legal recognition, both in cases and in statutes. *See,* for example, the Kansas statute defining death, reprinted in Jay Katz, *Experimentation with Human Beings* (New York, 1973), p. 1085, and also the discussion of cases, pp. 1076–1077, 1102–1104.

[17]For the presentation of a startling argument that we need to pronounce death in these cases precisely so we can *not* pull the plug but repeatedly "harvest" this new sort of corpse—for the blood it continues to produce, as an experimental object, as a training object for medical students, and so on, *see* Willard Gaylin, "Harvesting the Dead," *Harper's,* September 1974, 23–30.

[18]It has been argued that what I have called "giving up" should be regarded in law as a nonculpable *omission. See* George P. Fletcher, "Prolonging Life," Washington Law Review **42** (1967), 999. It is a persuasive argument.

[19]Consider the astonishing case reported by Beecher, *Research and the Individual,* no. **8,** p. 160.

A 5-year-old boy, for example, was submerged for 22 minutes in a Norwegian river at a temperature of –10°C. Before he went under, he was seen in the water clinging to the ice. Doubtless his body temperature rapidly fell during this period, and the resulting hypothermic state probably accounts for his survival. Although the boy seemed to be dead, with blue-white skin and widely dilated pupils, he was given mouth-to-mouth insufflation. The mouth and pharynx were filled with vomitus. This was partially cleared. No pulse was felt. The trachea was intubated, the airway aspirated, artificial respiration instituted, and external heart compression started at once and continued on the way to the hospital. On arrival, there was some evidence of peripheral circulation. The ear lobes became pink.

The heart was pricked with a needle and epinephrine and procaine were administered, without apparent result. Blood was withdrawn for typing and for determining the extent of hemolysis. Two-and-one-half hours after submersion, the heart started to contract spontaneously. Chlorpromazine was administered in an effort to improve the peripheral circulation. Gasping breaths now followed and soon became normal, but in an hour pulmonary edema appeared. Lanatoside and theophyllamine and morphine were given to control it. Three more pulmonary edema episodes ensued. An exchange transfusion was given to eliminate the free hemoglobin and potassium. Respiratory failure occurred five times in the next 24 hours. Hydrocortisone, antibiotics, heparin, and chlorpromazine were given. He was transfused. Examination two days after the accident showed no pupillary or corneal reflexes and no reaction to painful stimuli. On the fifth day, these signs returned. In a week, the boy began to swallow and to cough. On the tenth day he could obey simple commands, recognize his mother, and say "Yes," or "No." The next day he began to shriek and became restless and unconscious. Except for the brief period mentioned, he was unconscious for about six weeks. The agitated period lasted 14 days. He seemed to be decerebrated. Gradual improvement followed, but he appeared to be blind. Six weeks after the accident his mental condition improved. He began to speak, but still seemed to be blind. A week later, his vision began to return. On discharge, two-and-a-half months after the accident, he behaved like a normal child, except for a little ataxia. Six months after the accident, his mental condition was almost normal for his age, although he was still clumsy, and peripheral vision was reduced. Neurologic examination, including an electroencephalogram, was normal. By the usual clinical standards, he behaved as a normal child.

In Defense of Abortion and Infanticide

Michael Tooley

Introduction

This essay deals with the question of the morality of abortion and infanticide. The fundamental ethical objection traditionally advanced against these practices rests on the contention that human fetuses and infants have a right to life. It is this claim which will be the focus of attention here. The basic issue to be discussed, then, is what properties a thing must possess in order to have a right to life. My approach will be to set out and defend a basic moral principle specifying a condition an organism must satisfy if it is to have a right to life. It will be seen that this condition is not satisfied by human fetuses and infants, and thus that they do not have a right to life. So unless there are other objections to abortion and infanticide which are sound, one is forced to conclude that these practices are morally acceptable ones.[1] In contrast, it may turn out that our treatment of adult members of some other species is morally indefensible. For it is quite possible that some nonhuman animals do possess properties that endow them with a right to life.

Abortion and Infanticide

What reason is there for raising the question of the morality of infanticide? One reason is that it seems very difficult to formulate a completely satisfactory pro-abortion position without coming to grips with the infanticide issue. For the problem that the liberal on abortion encounters here is that of specifying a cutoff point which is not arbitrary: at what stage in the development of a human being does it cease to be morally permissible to destroy it, and why?

It is important to be clear about the difficulty here. The problem is not, as some have thought, that since there is a continuous line of development from a zygote to a newborn baby, one cannot hold that it is seriously wrong to destroy a newborn baby without also holding that it is seriously wrong to destroy a zygote, or any intermediate stage in the development of a human being. The problem is rather that if one says that it is wrong to destroy a newborn baby but not a zygote or some intermediate stage, one should be prepared to point to a *morally relevant* difference between a newborn baby and the earlier stage in the development of a human being.

Precisely the same difficulty can, of course, be raised for a person who holds that infanticide is morally permissible, since one can ask what morally relevant difference there is between an adult human being and a newborn baby. What makes it morally permissible to destroy a baby, but wrong to kill an adult? So the challenge remains. But I shall argue that in the latter case there is an extremely plausible answer.

Reflecting on the morality of infanticide forces one to face up to this challenge. In the case of abortion the number of events—quickening or viability, for instance—might be taken as cutoff points, and it is easy to overlook the fact that none of these events involves any morally significant change in the developing human. In contrast, if one is going to defend infanticide, one has to get very clear about what it is that gives something a right to life.

One of the interesting ways in which the abortion issue differs from most other moral issues is that the plausible positions on abortion appear to be extreme ones. For if a human fetus has a right to life, one is inclined to say that, in general, one would be justified in killing it only to save the life of the mother, and perhaps not even in that case.[2] Such is the extreme anti-abortion position. On the other hand, if the fetus does not have a right to life, why should it be seriously wrong to destroy it? Why would one need to point to special circumstances—such as the presence of genetic disease, or a threat to the woman's health—in order to justify such action? The upshot is that there does not appear to be any room for a moderate position on abortion as one finds, for example, in the Model Penal Code recommendations.[3]

Aside from the light it may shed on the abortion question, the issue of infanticide is both interesting and important in its own right.

The theoretical interest has been mentioned above: it forces one to face up to the question of what it is that gives something a right to life. The practical importance need not be labored. Most people would prefer to raise children who do not suffer from gross deformities or from severe physical, emotional, or intellectual handicaps. If it could be shown that there is no moral objection to infanticide, the happiness of society could be significantly and justifiably increased.

The suggestion that infanticide may be morally permissible is not an idea that many people are able to consider dispassionately. Even philosophers tend to react in a way that seems primarily visceral—offering no arguments and dismissing infanticide out of hand.

Some philosophers have argued, however, that such a reaction is not inappropriate, on the ground that, first, moral principles must, in the final analysis, be justified by reference to our moral feelings, or intuitions, and secondly, infanticide is one practice that is judged wrong by virtually everyone's moral intuition. I believe, however, that this line of thought is unsound, and I have argued elsewhere that even if one grants, at least for the sake of argument, that moral intuitions are the final court of appeal regarding the acceptability of moral principles, the question of the morality of infanticide is not one that can be settled by an appeal to our intuitions concerning it.[4] If infanticide is to be rejected, an argument is needed, and I believe that the considerations advanced in this essay show that it is unlikely that such an argument is forthcoming.

What Sort of Being Can Possess a Right to Life?

The issues of the morality of abortion and of infanticide seem to turn primarily upon the answers to the following four questions:

1. What properties, other than potentialities, give something a right to life?
2. Do the corresponding potentialities also endow something with a right to life?
3. If not, do they at least make it seriously wrong to destroy it?
4. At what point in its development does a member of the biologically defined species *Homo sapiens* first possess those nonpotential properties that give something a right to life?

The argument to be developed in the present section bears upon the answers to the first two questions.

How can one determine what properties endow a being with a right to life? An approach that I believe is very promising starts out from the observation that there appear to be two radically different sorts of reasons why an entity may lack a certain right. Compare, for example, the following two claims:

1. A child does not have a right to smoke;
2. A newspaper does not have a right not to be torn up.

The first claim raises a substantive moral issue. People might well disagree about it, and support their conflicting views by appealing to different moral theories. The second dispute, in contrast, seems an unlikely candidate for moral dispute. It is natural to say that newspapers just are not the sort of thing that can have any rights at all, including a right not to be torn up. So there is no need to appeal to a substantive moral theory to resolve the question whether a newspaper has a right not to be torn up.

One way of characterizing this difference, albeit one that will not especially commend itself to philosophers of a Quinean bent, is to say that the second claim, unlike the first, is true in virtue of a certain *conceptual* connection, and that is why no moral theory is needed in order to see that it is true. The explanation, then, of why it is that a newspaper does not have a right not to be torn up, is that there exists some property P such that, first, newspapers lack property P, and secondly, it is a conceptual truth that only things with property P can be possessors of rights.

What might property P be? A plausible answer, I believe, is set out and defended by Joel Feinberg in his paper, "The Rights of Animals and Unborn Generations."[5] It takes the form of what Feinberg refers to as the *interest principle*: "...the sorts of beings who *can* have rights are precisely those who have (or can have) interests."[6] And then, since "interests must be compounded somehow out of conations,"[7] it follows that things devoid of desires, such as newspapers, can have neither interests nor rights. Here, then, is one account of the difference in status between judgments such as (1) and (2) above.

Let us now consider the right to life. The interest principle tells us that an entity cannot have any rights at all, and *a fortiori*, cannot have a right to life, unless it is capable of having interests. This in itself may be a conclusion of considerable importance. Consider, for example, a fertilized human egg cell. Someday it will come to have desires and interests. As a zygote, however, it does not have desires, nor even the *capacity* for having desires. What about interests? This depends upon the account one offers of the relationship between desires and interests. It seems to me that a zygote cannot be properly spoken of as a subject of interests. My reason is roughly this. What is in a thing's interest is a function of its present and future desires, both those it will actually have and those it could have. In the case of an entity that is not presently capable of any desires, its interests must be based entirely upon the satisfaction of future desires. Then, since satisfaction of future desires presupposes the continued existence of the entity in question, anything which has an interest which is based upon the satisfaction of future desires must also have an interest in its own continued existence. Therefore, something that is not presently capable of having any desires at all—like a zygote—cannot have any interests at all unless it has interest in its own continued existence. I shall argue shortly, however, that a zygote cannot have such an interest. From this it will follow that it cannot have any interests at all, and this conclusion, together with the interest principle, entails that not all members of the species *Homo sapiens* have a right to life.

The interest principle involves, then, a thesis concerning a necessary condition which something must satisfy if it is to have a right to life, and it is a thesis which has important moral implications. It implies, for example, that abortions, if performed sufficiently early, do not involve any violation of a right to life. But on the other hand, the interest principle provides no help with the question of the moral status of human organisms once they have developed to the point where they do have desires, and thus are capable of having interests. The interest principle states that they *can* have rights. It does not state whether they *do* have rights—including, in particular, a right not to be destroyed.

It is possible, however, that the interest principle does not exhaust the connections between rights and interests. It formulates only a very general connection: a thing cannot have any rights at all unless it is capable of having at least some interest. May there not be more specific connections, between particular rights and particular sorts of interests? The following line of thought lends plausibility to this suggestion. Consider animals such as cats. Some philosophers are inclined to hold that animals such as cats do not have any rights at all. But let us assume, for the purpose of the present discussion, that cats do have some rights, such as a right not to be tortured, and consider the following claim:

   (3) A cat does not have a right to a university education. How is this statement to be regarded? In particular, is it comparable in status to the claim that children do not have a right to smoke, or, instead, to the claim that newspapers do not have a right to be torn up? To the latter, surely. Just as a newspaper is not the sort of thing that can have any rights at all, including a right not to be destroyed, so one is inclined to say that a cat, though it may have some rights, such as a right not to be tortured, is not the sort of thing that can possibly have a right to a university education.

   This intuitive judgment about the status of claims such as (3) is reinforced, moreover, if one turns to the question of the grounds of the interest principle. Consider, for example, the account offered by Feinberg, which he summarizes as follows:

> Now we can extract from our discussion of animal rights a crucial principle for tentative use in the resolution of the other riddles about the applicability of the concept of a right, namely, that the sorts of beings who *can* have rights are precisely those who have (or can have) interests. I have come to this tentative conclusion for two reasons: (1) because a right holder must be capable of being represented and it is impossible to represent a being that has no interests, and (2) because a right holder must be capable of being a beneficiary in his own person, and a being without interests is a being that is incapable of being harmed or benefitted, having no good or 'sake' of its own. Thus a being without interests has no 'behalf' to act in, and no 'sake' to act for.[8]

If this justification of the interest principle is sound, it can also be employed to support principles connecting particular rights with specific sorts of interests. Just as one cannot represent a being that has no interests at all, so one cannot, in demanding a university education for a cat, be representing the cat unless one is thereby representing some interest that the cat has, and that would be served by its receiving a university education. Similarily, one cannot be acting for the sake of a cat in arguing that it should receive a university education unless the cat has some interest that will thereby be furthered. The conclusion, therefore, is that if Feinberg's defense of the interest principle is sound, other, more specific principles must also be correct. These more specific principles can be summed up, albeit somewhat vaguely, by the following, *particular-interests principle*:

It is a conceptual truth that an entity cannot have a particular right, R, unless it is at least capable of having some interest, I, which is furthered by its having right R.

Given this particular-interests principle, certain familiar facts, whose importance has not often been appreciated, become comprehensible. Compare an act of killing a normal adult human being with an act of torturing one for five minutes. Though both acts are seriously wrong, they are not equally so. Here, as in most cases, to violate an individual's right to life is more seriously wrong than to violate his right not to have pain inflicted upon him. Consider, however, the corresponding actions in the case of a newborn kitten. Most people feel that it is seriously wrong to torture a kitten for five minutes, but not to kill it painlessly. How is this difference in the moral ordering of the two types of acts, between the human case and the kitten case, to be explained. One answer is that while normal adult human beings have both a right to life and a right not to be tortured, a kitten has only the latter. But why should this be so? The particular-interest principle, however, suggests a possible explanation. Though kittens have some interests, including, in particular, an interest in not being tortured, which derives from their capacity to feel pain, they do not have an interest in their own continued existence, and hence do not have a right not to be destroyed. This answer contains, of course, a large promissory element. One needs a defense of the view that kittens have no interest in continued existence. But the point here is simply that there is an important question about the rationale underlying the moral ordering of certain sorts of acts, and that the particular-interests principle points to a possible answer.

This fact lends further plausibility, I believe, to the particular-interests principle. What one would ultimately like to do, of course, is to set out an analysis of the concept of a right, show that the analysis is indeed satisfactory, and then show that the particular-interests principle is entailed by the analysis. Unfortunately, it will not be possible to pursue such an approach here, since formulating an acceptable analysis of the concept of a right is a far from trivial matter. What I should like to do, however, is to touch briefly upon the problem of providing such an analysis, and then to indicate the account that seems to me most satisfactory—an account which does entail the particular-interests principle.

It would be widely agreed, I believe, both that rights impose obligations, and that the obligations they impose upon others are *conditional* upon certain factors. The difficulty arises when one attempts to specify what the obligations are conditional upon. There seems to be two main views in this area. According to the one, rights impose obligations that are conditional upon the interests of the possessor of the right. To say that Sandra has a right to something is thus to say, roughly, that if it is in Sandra's interest to have that thing, then others are under an obligation not to deprive her of it. According to the second view, rights impose obligations that are conditional upon the right's not having been waived. To say that Sandra has a right to something is to say, roughly, that if Sandra has not given others permission to take the thing, then they are under an obligation not to deprive her of it.

Both views encounter serious difficulties. On the one hand, in the case of minors, and nonhuman animals, it would seem that the obligations that rights impose must be taken as conditional upon the interests of those individuals, rather than upon whether they have given one permission to do certain things. On the other, in the case of individuals who are capable of making informed and rational decisions, if that person has not given one permission to take something that belongs to him, it would seem that one is, in general, still under an obligation not to deprive him of it, even if having that thing is no longer in his interest.

As a result, it seems that a more complex account is needed of the factors upon which the obligations imposed by rights are conditional. The account which I now prefer, and which I have defended elsewhere[9], is this: "A has a right to X" means the same as:

> A is such that it can be in A's interest to have X, and either (1) A is not capable of making an informed and rational choice whether to grant others permission to deprive him of X, in which case, if it is in A's interest not to be deprived of X, then, by this fact alone, others are under a prima facie obligation not to deprive A of X, *or* (2) A is capable of making an informed and rational choice whether to grant others permission to deprive him of X, in which case the others are under a prima facie obligation not to deprive A of X if and only if A has not granted them permission to do so.

And if this account, or something rather similar is correct, then so is the particular-interests principle.

What I now want to do is simply apply the particular-interests principle to the case of the right to life. First, however, one needs to notice that the expression, "right to life," is not entirely happy, since it suggests that the right in question concerns the continued existence of a biological organism. That this is incorrect can be brought out by considering possible ways of violating an individual's right to life. Suppose, for example, that future technological developments make it possible to change completely the neural networks in a brain, and that the brain of some normal adult human being is thus completely reprogrammed, so that the organism in question winds up with memories (or rather, apparent memories), beliefs, attitudes, and personality traits totally different from those associated with it before it was subjected to reprogramming. (The Pope is reprogrammed, say, on the model of Bertrand Russell.) In such a case, however beneficial the change might be, one would surely want to say that *someone* had been destroyed, that an adult human being's right to life had been violated, even though no biological organism had been killed. This shows that the expression, "right to life," is misleading, since what one is concerned about is not just the continued existence of a biological organism.

How, then, might the right in question be more accurately described? A natural suggestion is that the expression, "right to life," refers to the right of a subject of experiences and other mental states to continue to exist. It might be contended, however, that this interpretation begs the question against certain possible views. For someone might hold—and surely some people in fact do—that while continuing subjects of experiences and other mental states certainly have a right to life, so do some other organisms that are only potentially such continuing subjects, such as human fetuses. A right to life, on this view, is *either* the right of a subject of experiences to continue to exist, *or* the right of something that is only potentially a continuing subject of experiences to become such an entity.

This view is, I believe, to be rejected, for at least two reasons. In the first place, this view appears to be clearly incompatible with the interest principle. Secondly, this position entails that the destruction of potential persons is, in general, *prima facie* seriously wrong, and I shall argue, in the next section, that the latter view is incorrect.

Let us consider, then, the right of a subject of experiences and other mental states to continue to exist. The particular-interests principle implies that something cannot possibly have such a right unless its continued existence can be in its interest. We need to ask, then, what must be the case if the continued existence of something is to be in its interest.

It will help to focus our thinking, I believe, if we consider a crucial case, stressed by Derek Parfit. Imagine a human baby that has developed to the point of being sentient, and of having simple desires, but that is not yet capable of having any desire for continued existence. Suppose, further, that the baby will enjoy a happy life, and will be glad that it was not destroyed. Can we or can we not say that it is in the baby's interest not to be destroyed?

To approach this case, let us consider a closely related one, namely, that of a human embryo that has not developed sufficiently far to have any desires, or even any states of consciousness at all, but that will develop into an individual who will enjoy a happy life, and who will be glad that his mother did not have an abortion. Can we or can we not say that it is the embryo's interest not to be destroyed?

Why might someone be tempted to say that it is in the embryo's interest that it not be destroyed? One line of thought which, I believe, tempts some people, is this. Let Mary be an individual who enjoys a happy life. Then, though some philosophers have expressed serious doubts about this, it might very well be said that it was certainly in Mary's interest that a certain embryo was not destroyed several years earlier. And this claim, together with the tendency to use expressions such as "Mary before she was born" to refer to the embryo in question, may lead one to think that it was in the embryo's interest not to be destroyed. But this way of thinking involves conceptual confusion. A subject of interests, in the relevant sense of "interest", must necessarily be a subject of conscious states, including experiences and desires. This means that in identifying Mary with the embryo, and attributing to it her interest in its earlier nondestruction, one is treating the embryo as if it were itself a subject of consciousness. But by hypothesis, the embryo being considered has not developed to the point where there is any subject of consciousness associated with it. It cannot, therefore, have any interests at all, and *a fortiori*, it cannot have any interest in its own continued existence.

Let us now return to the first case—that of a human baby that is sentient, and which has simple desires, but which is not yet capable of having more complex desires, such as a desire for its own continued existence. Given that it will develop into an individual who will lead a happy life, and who will be glad that the baby was not destroyed, does one want to say that the baby's not being destroyed is in the baby's own interest?

Again, the following line of thought may seem initially tempting. If Mary is the resulting individual, then it was in Mary's interest that the baby not have been destroyed. But the baby just *is* Mary when she was young. So it must have been in the baby's interest that it not have been destroyed.

Indeed, this argument is considerably more tempting in the present case than in the former, since here there is something that is a subject of consciousness, and which it is natural to identify with Mary. I suggest, however, that when one reflects upon the case, it becomes clear that such an identification is justified only if certain further things are the case. Thus, on the one hand, suppose that Mary is able to remember quite clearly some of the experiences that the baby enjoyed. Given that sort of causal and psychological connection, it would seem perfectly reasonable to hold that Mary and the baby are one and the same subject of consciousness, and thus, that if it is in Mary's interest that the baby not have been destroyed, then this must also have been in the baby's interest. On the other hand, suppose that not only does Mary, at a much later time, not remember any of the baby's experiences, but the experiences in question are not psychologically linked, either via memory or in any other way, to mental states enjoyed by the human organism in question at *any* later time. Here it seems to me clearly incorrect to say that Mary and the baby are one and the same subject of consciousness, and therefore it cannot be correct to transfer, from Mary to the baby, Mary's interest in the baby's not having been destroyed.

Let us now return to the question of what must be the case if the continued existence of something is to be in its own interest. The picture that emerges from the two cases just discussed is this. In the first place, nothing at all can be in an entity's interest unless it has desires at some time or other. But more than this is required if the continued existence of the entity is to be in its own interest. One possibility, which will generally be sufficient, is that the individual have, at the time in question, a desire for its own continued existence. Yet it also seems clear that an individual's continued existence can be in its own interest even when such a desire is not present. What is needed, apparently, is that the continued existence of the individual will make possible the satisfaction of some desires existing at other times. But not just any desires existing at other times will do. Indeed, as is illustrated both by the case of the baby just discussed, and by the deprogramming/reprogramming example, it is not even sufficient that they be desires associated with the same physical organism. It is crucial that they be desires that belong to one and the same subject of consciousness.

The critical question, then, concerns the conditions under which desires existing at different times can be correctly attributed to a single, continuing subject of consciousness. This question raises a number of difficult issues which cannot be considered here. Part of the rationale underlying the view I wish to advance will be clear, however, if one considers the role played by memory in the psychological unity of an individual over time. When I remember a past experience, what I know is not merely that there was a certain experience which someone or other had, but that there was an experience that belonged to the *same* individual as the present memory beliefs, and it seems clear that this feature of one's memories is, in general, a crucial part of what it is that makes one a continuing subject of experiences, rather than merely a series of psychologically isolated, momentary subjects of consciousness. This suggests something like the following principle:

> Desires existing at different times can belong to a single, continuing subject of consciousness only if that subject of consciousness possesses, at some time, the concept of a continuing self or mental substance.[10]

Given this principle, together with the particular-rights principle, one can set out the following argument in support of a claim concerning a necessary condition which an entity must satisfy if it is to have a right to life:

1. The concept of a right is such that an individual cannot have a right at time *t* to continued existence unless the individual is such that it can be in its interest at time *t* that it continue to exist.
2. The continued existence of a given subject of consciousness cannot be in that individual's interest at time *t* unless *either* that individual has a desire, at time *t*, to continue as a subject of consciousness, *or* that individual can have desires at other times.

3. An individual cannot have a desire to continue to exist as a subject of consciousness unless it possesses the concept of continuing self or mental substance.
4. An individual existing at one time cannot have desires at other times unless there is at least one time at which it possesses the concept of a continuing self or mental substance.

Therefore:

5. An individual cannot have a right to continued existence unless there is at least one time at which it possesses the concept of a continuing self or mental substance.

This conclusion is obviously significant. But precisely what implications does it have with respect to the morality of abortion and infanticide? The answer will depend upon what relationship there is between, on the one hand, the behavioral and neurophysiological development of a human being, and, on the other, the development of that individual's mind. Some people believe that there is no relationship at all. They believe that a human mind, with all its mature capacities, is present in a human from conception onward, and so is there before the brain has even begun to develop, and before the individual has begun to exhibit behavior expressive of higher mental functioning. Most philosophers, however, reject this view. They believe, on the one hand, that there is, in general, a rather close relation between an individual's behavioral capacities and its mental functioning, and, on the other, that there is a very intimate relationship between the mind and the brain. As regards the latter, some philosophers hold that the mind is in fact identical with the brain. Others maintain that the mind is distinct from the brain, but causally dependent upon it. In either case, the result is a view according to which the development of the mind and the brain are necessarily closely tied to one another.

If one does adopt the view that there is a close relation between the behavioral and neurophysiological development of a human being, and the development of its mind, then the above conclusion has a very important, and possibly decisive implication with respect to the morality of abortion and infanticide. For when human development, both behavioral and neurophysiological, is closely examined, it is seen to be most unlikely that human fetuses, or even newborn babies, possess any concept of a continuing self.[11] And in the light of the above conclusion, this means that such individuals do not possess a right to life.

But is it reasonable to hold that there is a close relation between human behavioral and neurophysiological development, and the development of the human mind? Approached from a scientific perspective, I believe that there is excellent reason for doing so. Consider, for example, what is known about how, at later stages, human mental capacities proceed in step with brain development, or what is known about how damage to different parts of the brain can affect, in different ways, an individual's intellectual capacities.

Why, then, do some people reject the view that there is a close relationship between the development of the human mind, and the behavioral and neurophysiological development of human beings? There are, I think, two main reasons. First, some philosophers believe that the scientific evidence is irrelevant, because they believe that it is possible to establish, by means of a purely metaphysical argument, that a human mind, with its mature capacities, is present in a human from conception onward. I have argued elsewhere that the argument in question is unsound.[12]

Secondly, and more commonly, some people appeal to the idea that it is a divinely revealed truth that human beings have minds from conception onward. There are a number of points to be made about such an appeal. In the first place, the belief that a mind, or soul, is infused into a human body at conception by God is not an essential belief within many of the worlds religions. Secondly, even with religious traditions, such as Roman Catholicism, where the belief is a very common one, it is by no means universally accepted. Thus, for example, the well-known Catholic philosopher, Joseph Donceel, has argued very strongly for the claim that the correct position on the question of ensoulment is that the soul enters the body only when the human brain has undergone a sufficient process of development.[13] Thirdly, there is the question of whether it is reasonable to accept the religious outlook which is being appealed to in support of the contention that humans have minds which are capable of higher intellectual activities from conception onward. This question raises very large issues in philosophy of religion which cannot be pursued here. But it should at least be said that most contemporary philosophers who have reflected upon religious beliefs have come to the view that there is not sufficient reason even for believing in the existence of God, let alone for accepting the much more detailed religious claims which are part of a religion such as Christianity. Finally, suppose that one nonetheless decides to accept the contention that it is a divinely revealed truth that humans have, from conception onward, minds that are capable of higher mental activities, and that one appeals to this purported revelation in order to support the claim that all humans have a right to life. One needs to notice that if one then goes on to argue, not merely that abortion is wrong, but that there should be a law against it, one will encounter a very serious objection. For it is surely true that it is inappropriate, at least in a pluralistic society, to appeal to specific religious beliefs of a nonmoral sort—such as the belief that God infuses souls into human bodies at conception—in support of legislation that will be binding upon everyone, including those who either accept different religious beliefs, or none at all.

Is it Morally Wrong to Destroy Potential Persons?

In this section I shall consider the question of whether it can be seriously wrong to destroy an entity, not because of the nonpotential properties it presently possesses, but because of the properties it will later come to have, if it is not interfered with. First, however, we need to be clear why this is such a crucial question. We can do this by considering a line of thought that has led some people to feel that the anti-abortionist position is more defensible than that of the pro-abortionist. The argument in question rests upon the gradual and continuous development of an organism as it changes from a zygote into an adult human being. The anti-abortionist can point to this development, and argue that it is morally arbitrary for a pro-abortionist to draw a line at some point in this continuous process—such as at birth, or viability—and to say that killing is permissible before, but not after, that particular point.

The pro-abortionist reply would be, I think, that the emphasis on the continuity of the process is misleading. What the anti-abortionist is really doing is simply challenging the pro-abortionist to specify what properties a thing must have in order to have a right to life, and to show that the developing organism does acquire those properties at the point in question. The pro-abortionist may then be tempted to argue that the difficulty in meeting this challenge should not

be taken as grounds for rejecting his position. For the anti -abortion-
ist cannot meet this challenge either; he is equally unable to say what
properties something must have if it is to have a right to life.

Although this rejoinder does not dispose of the anti-abortionist
argument, it is not without bite. For defenders of the view that abortion is almost always wrong have failed to face up to the question of
the *basic* moral principles on which their position rests—where a
basic moral principle is one whose acceptability does not rest upon
the truth of any factual claim of a nonmoral sort.[14] They have been
content to assert the wrongness of killing any organism, from a
zygote on, if that organism is a member of the biologically defined
species *Homo sapiens*. But they have overlooked the point that this
cannot be an acceptable *basic* moral principle, since difference in
species is not in itself a morally relevant difference.[15]

The anti-abortionist can reply, that it is possible to defend his
position, but not a pro-abortionist position, *without* getting clear
about the properties a thing must possess if it is to have a right to life.
For one can appeal to the following two claims. First, that there is a
property, even if one is unable to specify what it is, that (1) is
possessed by normal adult humans, and (2) endows any being possessing it with a right to life. Secondly, that there are properties
which satisfy (1) and (2), at least one of those properties will be such
that any organism potentially possessing that property has a right to
life even now, simply in virtue of that potentiality—where an
organism possesses a property potentially if it will come to have it in
the normal course of its development.

The second claim—which I shall refer to as the potentiality principle—is crucial to the anti-abortionist's defense of his position.
Given that principle, the anti-abortionist can defend his position
without grappling with the very difficult question of what nonpotential properties an entity must possess in order to have a right to life.
It is enough to know that adult members of *Homo sapiens* do have
such a right. For then one can employ the potentiality principle to
conclude that any organism which belongs to the species *Homo
sapiens*, from a zygote on—with the possible exception of those that
suffer from certain gross neurophysiological abnormalities—must
also have a right to life.

The pro-abortionist, in contrast, cannot mount a comparable argument. He cannot defend his position without offering at least a par-

tial answer to the question of what properties a thing must possess in order to have a right to life.

The importance of the potentiality principle, however, goes beyond the fact that it provides support for an anti-abortion position. For it seems that if the potentiality principle is unsound, then there is no acceptable defense of an extreme conservative view on abortion.

The reason is this. Suppose that the claim that an organism's having certain potentialities is sufficient grounds for its having a right to life cannot be sustained. The claim that a fetus which is a member of *Homo sapiens* has a right to life can then be attacked as follows. The reason an adult member of *Homo sapiens* has a right to life, but an infant ape, say, does not, is that there are certain physiological properties which the former possesses and the latter does not. Now even if one is unsure exactly what the relevant psychological characteristics are, it seems clear that an organism in the early stages of development from a zygote into an adult member of *Homo sapiens* does not possess those properties. One need merely compare a human fetus with an ape fetus. In early stages of development, neither will have any mental life at all. (Does a zygote have a mental life? Does it have experiences? Or beliefs? Or desires?) In later stages of fetal development some mental events presumably occur, but these will be of a very rudimentary sort. The crucial point, however, is that given what we know through comparative studies of, on the one hand, brain development, and, on the other behavior after birth, it is surely reasonable to hold that there are no significant differences in the respective mental lives of a human fetus and an ape fetus. There are, of course, physiological differences, but these are not in themselves morally significant. *If* one held that potentialities were relevant to the ascription of a right to life, one could argue that the physiological differences, though not morally relevant in themselves, are morally relevant in virtue of their causal consequences: they will lead to later psychological differences that are morally relevant, and for this reason the physiological differences are themselves morally significant. But if the potentiality principle is not available, this line of argument cannot be used, and there will then be no differences between a human and an ape fetus that the anti-abortionist can use as grounds for ascribing a right to life to the former but not to the latter.

This argument assumes, of course, that the anti-abortionist cannot successfully argue that there are religious reasons for holding that, even when potentialities are set aside, there is a morally relevant difference between human fetuses and ape fetuses. In the previous section I indicated, however, why it is very unlikely that any religious line of argument can be satisfactory in the present context.

The conclusion seems to be then, that the anti-abortionist position is defensible only if the potentiality principle is sound. Let us now consider what can be said against that principle. One way of attacking it is by appealing to the conclusion advanced in the previous section, to the effect that an individual cannot have a right to continued existence unless there is at least one time at which it possesses the concept of continuing self or mental substance. This principle entails the denial of the potentiality principle. Or more precisely, it does so in conjunction with the presumably uncontroversial empirical claim that a fertilized human egg cell, which does possess the relevant potentialities, does not possess the concept of a continuing self or mental substance.

Alternatively, one could appeal to the more modest claim involved in the interest principle, and use it to argue that since a fertilized human egg cell cannot have any interests at all, it cannot have any rights, and *a fortiori* cannot have a right to life. So potentialities alone cannot endow something with a right to life.

Given these lines of argument, is there any reason not to rest the case at this point? I want to suggest that there are at least two reasons why one needs to take a closer look at the potentiality principle. The first is that some people who are anti-abortionists may wish to reject not only the particular-interests principle, but also the more modest interest principle, and although I believe that this response to the above arguments is unsound, I think it is important to see whether there aren't other arguments that are untouched by this reply.

A second, and more important reason why it is unwise to base one's case against the anti-abortionist entirely upon an appeal to principles such as the interest principle is this. The anti-abortionist can modify his position slightly, and avoid the arguments in question. Specifically, he can abandon his claim that a human fetus has a right to life, but contend that it is nevertheless seriously wrong to kill it. Some philosophers would feel that such a modification cannot possibly be acceptable, on the ground that no action can be seriously wrong unless it violates someone's right to something. It seems to me, however, that this latter view is in fact mistaken.[16] In any case, let us consider the position that results from this modification. An anti-abortionist who is willing to adopt this position can then appeal, not to the potentiality principle, but to the following, *modified potentiality principle:*

> If there are properties possessed by normal adult human beings that endow any organism possessing them with a right to life, then at least one of those properties is such that it is seriously wrong to kill any organism that potentially possesses that property, simply in virtue of that potentiality.

Since this modified potentiality principle is not concerned with the attribution of rights to organisms, it cannot be attacked by appealing to the interest principle, or to the particular-interests principle, or to some analysis of the concept of a right.

Let us now consider how the case against the anti-abortionist position can be strengthened. I shall advance three arguments which are objections to both the original and the modified potentiality principles. Since the original potentiality principle cannot be correct unless the modified one is, it will suffice to consider only the modified principle. The basic issue, then, is: Is there any property J which satisfies the following three conditions:

1. There is a property, K, such that any individual possessing property K has a right to life, and there is a scientific law, L, to the effect that any organism possessing property J will, in the normal course of events, come to possess property K at some later time;
2. Given the relationship just described between property J and property K, it is seriously wrong to kill anything possessing property J;
3. If property J were not related to property K in the way indicated the fact that an organism possessed property J would not make it seriously wrong to kill it.

In short, the question is whether there is a property, J, that makes it seriously wrong to kill something *only because* J stands in a certain causal relation to a second property, K, which is such that anything possessing that property *ipso facto* has a right to life.

My first objection turns upon the claim that if one accepts the modified potentiality principle, one ought also to accept the following, *generalized potentiality principle:*

> If there are any properties possessed by normal adult human beings that endow any organism possessing them with a right to life, then at

least one of those properties is such that it is seriously wrong to perform any action that will prevent some system, which otherwise
would have developed the property, from doing so.

This generalized potentiality principle differs from the original and
the modified potentiality principles in two respects. First, it applies
to *systems* of objects, and not merely to organisms. I think that this
first generalization is one that ought to be accepted by anyone who
accepts either the original or the modified principle. For why should
it make any difference whether the potentiality resides in a single organism, or in a system of organisms that are so interrelated that they
will in the normal course of affairs, due to the operation of natural
laws, causally give rise to something that possesses the property in
question? Surely it is only the potentiality for a certain outcome that
matters, and not whether there are one or more objects interacting
and developing in a predetermined way to produce that outcome.

In thinking about this issue, it is important not to confuse *potentialities* with mere *possibilities*. The generalized potentiality principle does not deal with collection of objects that merely have the
capacity to interact in certain ways. The objects must already be
interrelated in such a way that in the absence of external interference
the laws governing their future interaction and development will
bring it about that the system will develop the property in question.

The second difference is that the original and modified potentiality principles deal only with the *destruction* of organisms, while the
generalized principle deals with any action that prevents an organism, or a system, from developing the relevant property. I think that
the anti-abortionist will certainly want to accept this generalization.
For suppose that by exposing a human zygote to appropriate radia-
tion one could transform it into a frog zygote. A woman could then
undergo a two-step abortion: first the human zygote would be transformed into a frog zygote, and then the frog zygote would be destroyed. Assuming that one does not view the destruction of a frog
zygote as seriously wrong, one must either hold that it is seriously
wrong to prevent a human zygote from developing its potentialities,
or else conclude that the two-step abortion technique is morally permissible. The latter option would not appear to be a viable, let alone
a welcome one, for the anti-abortionist. For why should it be morally permissible to destroy a human organism in two steps, one of

which limits its potentialities, and the other which destroys the resulting organism, but seriously wrong to collapse these two steps into one, limiting its potentialities and destroying it by a single action? I think that the anti-abortionists would agree that there is no significant moral distinction here, and thus would accept the second generalization involved in the expanded potentiality principle.

Suppose, now, that artificial wombs have been perfected. A healthy, unfertilized human egg cell has been placed in one, along with a large number of spermatozoa. If the device is turned on, the spermatozoa will be carried, via a conveyor belt, to the unfertilized egg cell, where, we can assume, fertilization will take place. The device is such, moreover, that no outside assistance will be needed at any future stage, and nine months later a normal human baby will emerge from the artificial womb. Given these assumptions—all of which are certainly empirically possible—once such a device has been turned on, there will exist an active potentiality that will, if not interfered with, give rise to something that will become an adult human being, and so will have a right to life. But would it be seriously wrong to destroy that potentiality—as might be done, for example, by turning off the machine, or by cutting the conveyor belt, so that fertilization does not take place? Most people, I believe, would certainly not think that such actions were seriously wrong. If that view is correct, the generalized potentiality principle must be rejected as unsound.

In short, the first argument against the modified potentiality principle, and hence against the original potentiality principle, is as follows. It is reasonable to accept the modified principle only if it is also reasonable to accept the generalized potentiality principle, because whether the potentialities reside in a single organism or in a system does not seem to be a morally significant difference. But to accept the generalized potentiality principle is to commit oneself to the view that interference with an artificial womb so as to prevent fertilization from taking place is just as seriously wrong as abortion, and for precisely the same reason. If, as seems plausible, this is not an acceptable view, then one cannot reasonably accept either the original or the modified potentiality principle.

Let us now turn to my second argument against the modified potentiality principle. This argument turns upon the following crucial claim:

Let C be any type of causal process where there is some type of occurrence, E, such that processes of type C would possess no intrinsic moral significance were it not for the fact that they result in occurrences of type E.

Then:

The characteristic of being an act of intervening in a process of type C which prevents the occurrence of an outcome of type E makes an action intrinsically wrong to precisely the same degree as does the characteristic of being an act of ensuring that a causal process of type C, which it was in one's power to initiate, does not get initiated.

This principle, which I shall refer to as the moral symmetry principle with respect to action, would be rejected by some philosophers. They would argue that there is an important distinction to be drawn between "what we owe people in the form of aid, and what we owe them in the way of noninterference,"[17] and that the latter, "negative duties," are duties that it is more serious to neglect than the former, "positive" ones. This view arises from an intuitive response to examples such as the following. Even if it is wrong not to send food to starving people in other parts of the world, it is more wrong still to kill someone. And isn't the conclusion, then, that one's obligation to refrain from killing someone is a more serious obligation than one's obligation to save lives?

I want to argue that this is not the correct conclusion. I think that it is tempting to draw this conclusion if one fails to consider the motivation likely to be associated with the respective actions. If someone performs an action he knows will kill someone else, this will usually be good reason for concluding that he wanted the death of the person in question. In contrast, failing to help someone may indicate only apathy, laziness, selfishness, or an amoral outlook: the fact that a person knowingly allows another to die is not normally grounds for concluding that he desired that person's death. Someone who knowingly kills another is thus more likely to be seriously defective from a moral point of view than someone who fails to save another's life.

If we are not to be led to false conclusions by our intuitions about certain cases, we must explicitly assume identical motivations in the two situations. Compare, for example, the following: (1) Jones sees that Smith will be killed by a bomb unless he warns him. Jones' reaction is: "How fortunate, this will save me the trouble of killing

Smith myself." So Jones allows Smith to be killed by the bomb, even though he could have easily warned him. (2) Jones wants Smith dead, and therefore shoots him. Is one to say that there is a significant difference between the wrongness of Jones' behavior in these two cases? I suggest that it is not plausible to hold that there is a significant difference here.

If this is right, then it would appear to be a mistake to draw a distinction between positive and negative duties, and to hold the latter impose stricter obligations than the former. The differences in our intuitions about situations that involve giving aid to others, and corresponding situations that involve not interfering with others, is to be explained by reference to probable differences in the motivations that are likely to be present in the two sorts of situations, and not by reference to a distinction between positive and negative duties. For once it is specified that the motivation is the same in the two situations, it seems clear that inaction is as wrong in the one case as action is in the other.

There is another point that may be relevant. Action involves effort, whereas inaction usually does not. It does not usually require any effort to refrain from killing someone, but saving someone's life may require a considerable effort. One needs to ask, then, how large a sacrifice a person is morally required to make to save the life of another. If the sacrifice is a very significant one, it may be that one is not morally obliged to save the life of another in that situation. Superficial reflection upon such cases might easily lead one to introduce the distinction between postitive and negative duties, but again it seems clear that this would be a mistake. The point is not that one has a greater duty to refrain from killing others than to perform actions that will save them. It is rather that positive actions require effort, and this means that in deciding what to do a person has to take into account his own right to do what he wants with his life, and not only the other person's right to life. In order to avoid this confusion, then, one needs to confine oneself to consideration of situations in which the positive action requires no greater sacrifice than is involved in the inaction.

What I have been arguing, in brief, is this. It is probably true, for example, that most cases of killing are morally worse than most cases of merely letting die. This, however, is not an objection to the moral symmetry principle, since that principle does not imply that, all things considered, acts of killing are, in general, morally on a par with cases of allowing someone to die. What the moral symmetry principle implies is rather that, *other things being equal*, it is just as wrong to fail to save someone as it is to kill someone. If one wants to test this principle against one's moral intuitions, one has to be careful to select pairs of situations in which all other morally relevant factors—such as motivation, and risk to the agent—are equivalent. And I have suggested that when this is done, the moral symmetry principle is by no means counterintuitive.[18]

My argument against the modified potentiality principle can now be stated. Suppose at some future time a chemical were to be discovered which when injected into the brain of a kitten would cause the kitten to develop into a cat possessing a brain of the sort possessed by humans, and consequently into a cat having all the psychological capabilities characteristic of normal adult humans. Such cats would be able to think, to use language, and so on. Now it would surely be morally indefensible in such a situation to hold that it is seriously wrong to kill an adult member of the species *Homo sapiens* without also holding that it is wrong to kill any cat that has undergone such a process of development: there would be no morally significant differences.

Secondly, imagine that one has two kittens, one of which has been injected with a special chemical, but which has not yet developed those properties that in themselves endow something with a right to life, and the other of which has not been injected with the special chemical. It follows from the moral symmetry principle that the action of injecting the former with a "neutralizing" chemical that will interfere with the transformation process and prevent the kitten from developing those properties that in themselves would give it a right to life is *prima facie* no more seriously wrong than the action of intentionally refraining from injecting the second kitten with the special chemical.

It perhaps needs to be emphasized here that the moral symmetry principle does not imply that neither action is morally wrong. Perhaps both actions are wrong, even seriously so. The moral symmetry principle implies only that if they are wrong, they are so to precisely the same degree.

Thirdly, compare a kitten that has been injected with the special chemical and then had it neutralized, with a kitten that has never been injected with the chemical. It is clear that it is no more seriously wrong to kill the former than to kill the latter. For although their bodies have undergone different processes in the past, there is no reason why the kittens need differ in any way with respect to either their present properties or their potentialities.

Fourthly, again consider two kittens, one of which has been injected with the special chemical, but which has not yet developed those properties that in themselves would give it a right to life, and the other of which has not been injected with the chemical. It follows from the previous two steps in the argument that the combined action of injecting the first kitten with a neutralizing chemical and then killing it is no more seriously wrong than the combined action of intentionally refraining from injecting the second kitten with the special chemical and then killing it.

Fifthly, one way of neutralizing the action of the special chemical is simply to kill the kitten. And since there is surely no reason to hold that it is more seriously wrong to neutralize the chemical and to kill the kitten in a single step than in two successive steps, it must be the case that it is no more seriously wrong to kill a kitten that has been injected with the special chemical, but which has not developed those properties that in themselves would give it a right to life, than it is to inject such a kitten with a neutralizing chemical and then to kill it.

Next, compare a member of *Homo sapiens* that has not developed far enough to have those properties that in themselves give something a right to life, but which later will come to have them, with a kitten that has been injected with the special chemical but which has not yet had the chance to develop the relevant properties. It is clear that it cannot be any more seriously wrong to kill the human than to kill the kitten. The potentialities are the same in both cases. The only difference is that in the case of a human fetus the potentialities have been present from the beginning of the organism's development, while in the case of the kitten they have been present only from the time it was injected with the special chemical. This difference in the time at which the potentialities were acquired is not a morally relevant one.

It follows from the previous three steps in the argument that it is no more seriously wrong to kill a human being that lacks properties that in themselves, and irrespective of their causal consequences, endow something with a right to life, but which will naturally develop those properties, than it would be to intentionally refrain from injecting a kitten with the special chemical, and to kill it. But if it is the case that normal adult humans do possess properties that in themselves give them a right to life, it follows in virtue of the modified potentiality principle that it is seriously wrong to kill any human organism that will naturally develop the properties in question. Thus, if the modified potentiality principle is sound, we are forced by the above line of argument to conclude that if there were a chemical that would transform kittens into animals having the psychological capabilities possessed by adult humans, it would be seriously wrong to intentionally refrain from injecting kittens with the chemical, and to kill them instead.

But is it clear that this final conclusion is unacceptable? I believe that it is. It turns out, however, that this issue is *much* more complex than most people take it to be.[19] Here, however, it will have to suffice to note that the vast majority of people would certainly view this conclusion as unacceptable. For while there are at present no special chemicals that will transform kittens in the required way, there are other biological organisms, namely unfertilized human egg cells, and special chemicals, namely human spermatazoa, that will transform those organisms in the required way. So if one were to hold that it was seriously wrong to intentionally refrain from injecting kittens with the special chemical, and instead to kill them, one would also have to maintain that it was *prima facie* seriously wrong to refrain from injecting human egg cells with spermatozoa, and instead to kill them. So unless the anti-abortionist is prepared to hold that any woman, married or unmarried, does something seriously wrong every month that she intentionally refrains from getting pregnant, he cannot maintain that it would be seriously wrong to refrain from injecting the kitten with the special chemical, and instead to kill it.

In short, the above argument shows that anyone who wants to defend the original or the modified potentiality principle must either argue against the moral symmetry principle, or hold that in a world in which kittens could be transformed into "rational animals", it would be seriously wrong to kill newborn kittens. But we have just seen that if one accepts the latter claim, one must also hold that it is seriously wrong to intentionally refrain from fertilizing a human egg cell, and to kill it instead. Consequently, it seems very likely that any anti-abortionist rejoinder to the present argument will be directed against the moral symmetry principle. In the present essay I have not attempted to offer a thorough defense of that principle, although I have tried to show that what is perhaps the most important objection to it—the one that appeals to a distinction between positive and negative duties—rests upon a superficial analysis of our moral intuitions. Elsewhere, however, I have argued that a thorough examination of the moral symmetry principle sustains the conclusion that that principle is in fact correct.

There is one final point that needs to be made about the present argument, and that is that there are variations of it which do not involve the moral symmetry principle, but which also tell against the anti-abortionist position. These variations are suggested by the reflection that someone not quite convinced, for example, that failing to save is, in itself, just as seriously wrong as killing, may very well accept one of the following, more modest claims:

1. Failing to save someone is, in itself, almost as seriously wrong as killing.
2. Killing and failing to save are *comparable* in the technical sense of there being some number n such that failing to save n people is, in itself, at least as seriously wrong as killing one person.

These claims can be generalized into principles similar to, but less controversial than, the moral symmetry principle. In the case of the second we have the following, *moral comparability principle*:

> Let C be any type of causal process where there is some type of occurrence, E, such that processes of type C would possess no intrinsic moral significance were it not for the fact that they result in occurrences of type E.

Then:

> There is some number n such that the characteristic of being an act of ensuring that n causal processes of type C, which it was within one's power to initiate, do not get initiated, makes an action intrinsically wrong to at least the degree that the characteristic of being an act of intervening in a process of type C which prevents the occurrence of an event of type E does.

Given this moral comparability principle, one way of proceeding is simply to parallel the argument involving the moral symmetry principle. This will lead to the conclusion that there is some number n such that intentionally refraining from fertilizing n human egg cells is in itself at least as seriously wrong as destroying a human before it has acquired those nonpotential properties that would give it a right to life. Then, if it is granted that it is not seriously wrong to refrain from fertilizing n human egg cells, it follows that it is not in itself seriously wrong to destroy a human before it has acquired nonpotential properties that would give it a right to life.

Most people would readily grant the claim that it is not seriously wrong to intentionally refrain from fertilizing n human egg cells. It is worth noting, however, that there is a variant of the argument that avoids this assumption. Consider the moral status of destroying a human organism that has not yet acquired the nonpotential properties that would give it a right to life, and that, if allowed to survive, will lead an unhappy life. The moral comparability principle implies that there is some number n such that the destruction of such an entity is, in itself, no more seriously wrong than intentionally refraining from fertilizing n human egg cells that will result in n individuals who will lead unhappy lives. Since the latter is surely not even wrong, let alone seriously so, it follows that it is not seriously wrong to destroy a human organism that has not yet acquired those nonpotential properties that would give it a right to life, and that, if allowed to survive, will lead an unhappy life.

This variation of the comparability argument has led to a more modest conclusion. However, it is still a conclusion that tells against the view advanced by anti-abortionists, since they reject the claim that abortion is morally permissible in cases where the resulting individual will lead an unhappy life.

To sum up, what I have argued in the present section is this. The anti-abortionist position is defensible only if some version of the potentiality principle is sound. The original version of that principle is incompatible, however, both with the particular-interests principle and with the interest principle, and also with the account of the concept of a right offered above. The modified potentiality principle avoids these problems. There are, however, at least three other serious objections which tell against both the original potentiality principle and the modified one. It would seem, therefore, that there are excellent reasons for rejecting the potentiality principle, and with it, the anti-abortionist position.

Summary and Conclusions

In this paper I have advanced three main philosophical contentions:

1. An entity cannot have a right to life unless it is capable of having an interest in its own continued existence;
2. An entity is not capable of having an interest in its own continued existence unless it possesses, at some time, the concept of a continuing self, or subject of experiences and other mental states;
3. The fact that an entity will, if not destroyed, come to have proper ties that would give it a right to life does not in itself make it seriously wrong to destroy it.

If these philosophical contentions are correct, the crucial question is a factual one: At what point does a developing human being acquire the concept of a continuing self, and at what point is it capable of having an interest in its own continued existence? I have not examined this issue in detail here, but I have suggested that careful scientific studies of human development, both behavioral and neurophysiological, strongly support the view that even newborn humans do not have the capacities in question. If this is right, then it would seem that infanticide during a time interval shortly after birth must be viewed as morally acceptable.

But where is the line to be drawn? And what is the precise cutoff point? If one maintained, as some philosophers do, that an individual can possess a concept only if it is capable of expressing that concept linguistically, then it would be a relatively simple matter to determine whether a given organism possessed the concept of a continuing subject of experiences and other mental states. It is far from clear, however, that this claim about the necessary connection between the possession of concepts and the having of linguistic capabilities is correct. I would argue, for example, that one wants to ascribe mental states of a conceptual sort—such as beliefs and desires—to animals that are incapable of learning a language, and that an individual cannot have beliefs and desires unless it possesses the concepts involved in those beliefs and desires. And if that view is

right—if an organism can acquire concepts without thereby acquiring a way of expressing those concepts linguistically—then the
question of whether an individual possesses the concept of a continuing self may be one that requires quite subtle experimental techniques to answer.

If this view of the matter is roughly correct, there are two worries
that one is left with at the level of practical moral decisions, one of
which may turn out to be deeply disturbing. The lesser worry is the
question just raised: Where is the line to be drawn in the case of
infanticide? This is not really a troubling question since there is no
serious need to know the exact point at which a human infant
acquires a right to life. For in the vast majority of cases in which
infanticide is desirable due to serious defects from which the baby
suffers, its desirability will be apparent at birth or within a very short
time thereafter. Since it seems clear that an infant at this point in its
development is not capable of possessing the concept of a continuing subject of experiences and other mental states, and so is incapable of having an interest in its own continued existence, infanticide
will be morally permissible in the vast majority of cases in which it
is, for one reason or another, desirable. The practical moral problem
can thus be satisfactorily handled by choosing some short period of
time, such as a week after birth, as the interval during which infanticide will be permitted.

The troubling issue that arises out of the above reflections concerns whether adult animals belonging to species other than *Homo
sapiens* may not also possess a right to life. For once one allows that
an individual can possess concepts, and have beliefs and desires,
without being able to express those concepts, or those beliefs and
desires, linguistically, then it becomes very much an open question
whether animals belonging to other species do not possess properties that give them a right to life. Indeed, I am strongly inclined to
think that adult members of at least some nonhuman species do have
a right to life. My reason is that, first, I believe that some nonhuman
animals are capable of envisaging a future for themselves, and of
having desires about future states of themselves. Secondly, that anything which exercises these capacities has an interest in its own continued existence. And thirdly, that having an interest in one's own
continued existence is not merely necessary, but also a sufficient
condition, for having a right to life.

The suggestion that at least some nonhuman animals have a right to life is not unfamiliar, but is one that most of us are accustomed to dismissing very casually. The line of thought advanced here suggests that this attitude may very well turn out to be tragically mistaken. Once one reflects upon the question of the *basic* moral principles involved in the ascription of a right to life to organisms, one may find oneself driven to the conclusion that our everyday treatment of members of other species is morally indefensible, and that we are in fact murdering innocent persons.

Notes and References

[1]My book, *Abortion and Infanticide* (Oxford University Press, 1983), contains a detailed examination of other important objections.

[2]Judith Jarvis Thomson, in her article "A Defense of Abortion," *Philosophy & Public Affairs*, 1, 1971, 47–66, argues very forcefully for the view that this conclusion is incorrect. For a critical discussion of her argument, *see* Chapter 3 of *Abortion and Infanticide*.

[3]American Law Institute, *Model Penal Code* (Philadelphia, 1962), section 230.3.

[4]*Abortion and Infanticide*, chapter 10.

[5]In *Philosophy and Environmental Crisis*, William T. Blackstone ed., (Athens, Georgia, 1974), pp. 43–68.

[6]*Op. cit.*, p. 51.

[7]*Ibid.*, pp. 49–50.

[8]*Ibid.*, p. 51.

[9]*Op. cit.*, section 5.2.

[10]For a fuller discussion, and defense of this principle, *see op. cit.*, section 5.3.

[11]For a detailed survey of the scientific evidence concerning human development, *see op. cit.*, section 11.5.

[12]*Op. cit.*, section 11.42.

[13]For a brief discussion, see Joseph F. Donceel, "A Liberal Catholic's View," *Abortion in a Changing World*, vol. 1, edited by R. E. Hall (New York, 1970). A more detailed philosophical discussion can be found in Donceel's "Immediate Animation and delayed Hominization," *Theological Studies*, 31, 1970, 76–105.

[14]Consider the belief that it is *prima facie* wrong to pull cats' tails. Here is a belief that is almost universally accepted, but very few people, if any, would regard it as a basic moral belief. For this belief rests upon a nonmoral belief, to the effect that pulling cats' tails causes them pain. If one came to believe that cats actually enjoy this, one would abandon the moral belief in question. So the belief, though widely and firmly accepted, is a derived moral belief, rather than a basic one.

[15]For a much more extended discussion of this point, see, for example, Peter Singer's essay, "Animals and the Value of Life," *Matters of Life and Death*, ed. by

Tom Regan (Philadelphia, 1980), or my own discussion in section 4.2 of *Abortion and Infanticide*.

[16]*Op. cit.*, section 7.33.

[17]Philippa Foot, "The Problem of Abortion and the Doctrine of the Double Effect," *The Oxford Review*, **5**, 1967, 5–15. *See* the discussion of pp. 11ff.

[18]For a much more detailed defense of this view, see section 6.5 of *Abortion and Infanticide*.

[19]A discussion of why this is so can be found in Chapter 7 of *Abortion and Infanticide*.

The Perils of Personhood

Roslyn Weiss

In the abortion debate, one of the more overworked arguments concerns the if and when of the humanity of a fetus. For those who argue along these lines, the answer to these questions guarantees a resolution of the entire abortion issue. Thus, if the fetus is human, it must not be aborted except when the mother's life is endangered (and, for some, not even then): but if it is not human, it may be aborted under any circumstances.

Despite some fine discussions taking this approach,[1] many sense its futility, if only because each side is hopelessly deaf to the protestations of the other; each insists not merely on the correctness of its view but on its self-evidence.[2] Inevitably, philosophers turn elsewhere for answers.

One alternative is to challenge the validity of the implication. "If X is human then X may not be killed" (instead of focusing on the truth of the antecedent). This is the route traveled by Thomson. In her article, "A Defense of Abortion,"[3] Thomson grants the humanity of the fetus from the moment of conception, but nevertheless rejects the view that abortion is therefore impermissible in all cases (notably in the case of rape).

Another alternative, supported by Mary Anne Warren and Michael Tooley,[4] among others, diverts our attention from such concepts as 'human,' 'human being,' and 'Homo sapiens' as they apply to the fetus, and centers it instead on a new concept, 'personhood,' claimed to be distinct from these others in a most crucial and decisive way. It is to this concept that I now turn, as the major concern of this paper.

I intend, then, to propose and defend the following three theses:

I. 'Personhood' is the natural and logical consequence of a rights-centered approach to abortion.

II. The implications of the 'personhood' view are at times absurd and at others dangerous.

III. Rights alone can therefore not form a solid basis for the moral issue of abortion: The importance of duties emerges.

'Personhood' and the Role of Rights in the Abortion Issue

Ordinary usage, for the most part, easily interchanges such expressions as 'human being' and 'person' (except perhaps in these modern times when 'person' becomes synonymous with 'woman' as in 'salesperson,' 'chairperson,' and spokesperson.'), and, until recently, philosophers have followed suit.[5] But something in the abortion debate has caused some to press for a hard-and-fast line between these concepts. Our first task, then, will be to discover just what caused this change.

Let us begin by considering this fairly standard statement of the value of defining 'humanity': "The importance of these definitional questions for moral philosophy is obvious. Human beings protect themselves with a thicket or rights they do not grant to other beings, and some of these rights are said to be *human* rights—rights one has simply by virtue of being human. Any conceptual uncertainty about when an entity has become or has ceased to be human is a problem for the ascription of such rights."[6]

The implication of this excerpt is clear: Considerations regarding the definition and application of the term 'human' are not essential in themselves, but as a means to a further end; the *end* is plainly the ascription of rights. We ask whether or not the fetus is human because we want to know whether or not it has rights. The attempt to determine the moral permissibility or impermissibility of abortion by way of defining 'humanity' is thus ultimately grounded in the notion of rights.

Once the appeal to the definition of humanity is revealed as a concern about rights, the following natural, logical, inescapable question arises: If it is rights we are after, why seek definitions of humanity? Unless there is some *necessary* connection between being *Homo sapiens* and having rights, it is far more productive (because it assumes much less) to tackle directly *this* question: What sort of thing has rights? This is a moral question—not a question of biological fact—and hence far more appropriate in this context. For, as Tooley notes, difference in species is not in itself a morally relevant difference, and since what we seek *is* a morally relevant difference, our distinction must be drawn not between 'humans' and 'nonhumans' but rather between entities that have rights and entities that do not, between 'persons' and 'nonpersons.'

It must be stressed that the concept of personhood is not introduced as a contribution to the debate on the humanness of the fetus; it is not a criterion for determining if and when a thing becomes human.[7] For proponents of personhood, discussions of the humanity of the fetus are *at best* arbitrary and irrelevant; a brief survey of the host of stages in a fetus's development proposed as possible boundaries between nonhuman and human (conception, spontaneous motility, quickening, appearance of hands and feet, viability, birth) makes this all too apparent. At worst, these discussions are biased and self-serving, for they presuppose that humans, *and only humans*, have rights. What is called for, therefore, is a nonarbitrary, rationally defensible, morally relevant standard by which to establish not the biological status of the fetus but its moral status. We must seek criteria relevant to a thing's having rights, that is, to its personhood, and must steer quite clear of any attempt to find criteria by which to determine a thing's humanity.

Indeed, when our goal is to ascertain whether or not a fetus has rights, the question, "Is X a person?" marks an improvement over the question, "Is X human?" for two reasons: (1) It escapes hopeless entanglement in the numerous arbitratry and irrelevant points of entry into humanity, and (2) it eradicates the 'speciesist'[8] bias that only *Homo sapiens* can have rights.

The Dangerous and Absurd Implications of the Personhood View

Although the concept of personhood has definite advantages over that of humanity, it is not entirely problem free. Its concentration exclusively on rights renders it subject to its own brand of arbitrariness. "Rights are the stamping ground of intuitionists," says Hare,[9] and no means of assigning rights is fully immune to the intrinsic subjectivity of intuition.

Warren certainly seems to be aware of this, for she admits that there are likely to be a great many problems in formulating even rough criteria for personhood (i.e., for having rights). But she does propose a tentative formulation of five such criteria (consciousness, and in particular the capacity to feel pain; reasoning; self-motivated activity; the capacity to communicate; the presence of self-concepts, and self-awareness, either individual or racial, or both), along with two important qualifiers: (1) that an entity does not need to have all these attributes to be properly considered a person, and (2) that it is not absolutely essential that any one of these conditions be a necessary one. However, she does think that any being which satisfies none of these criteria is certainly not a person.

Tooley, on the other hand, refuses to "share the general pessimism about the possibility of resolving the issue of abortion" in terms of rights,[10] for he believes himself to have found a satisfactory and *nonarbitrary* necessary condition for having rights: for him, it is quite evident that if entity A either lacks consciousness, or has consciousness but is incapable of desiring X, then A has no right to X.

Interestingly, neither Warren nor Tooley explicitly specifies sufficient conditions for personhood; for their purposes, they simply do not require sufficient conditions. Since in their hands, the criteria for personhood are primarily weapons in the fight to legalize and vindicate voluntary abortion (and quite effective ones at that, since they never dirty those hands by denying the fetus's humanity), [11] all they wish to do is disqualify fetuses as persons. For this limited end, it is surely sufficient to show that fetuses do not fulfill some *necessary* condition. It is quite evident that the fetus fulfills neither Warren's nor Tooley's criteria for personhood.

One unfortunate side effect of this emphasis on personhood is that since infants are no more (or scarcely more) persons than are fetuses,[12] they too must be denied rights. Appropriately, this is the position of both Warren and Tooley,[13] and Tooley is even hopeful that infanticide may, in some cases, increase human happiness: with infanticide, people would not be forced to raise children who suffer from gross deformities or from severe physical, emotional, or intellectual handicaps. The strong emotions that infanticide arouses in us, he claims, are visceral reactions—not rational ones—and, as such, surely should not be allowed to stand in the way of increased human happiness.

Of course, there is no reason to assume that only fetuses and infants are to be excluded from personhood. Especially when we are seeking to increase human happiness, it becomes of the utmost importance to check the personhood of *anyone* who poses a threat to such happiness. In this context, we begin to appreciate the importance of sufficient conditions. As Tooley himself is aware, his necessary (but not necessarily sufficient) condition may be a theoretically incorrect cut off point, but if it is, at least it has the merit or erring on the side of caution. But if we are concerned with maximizing human happiness, caution is no virtue: we should aim at accuracy. We should make every effort to discover both the necessary and the sufficient conditions for personhood, and permit all nonpersons to be killed. A suitable necessary and sufficient condition for personhood might be the ability to communicate, for surely if one can claim rights, one can have them.[14] But even if we regard Tooley's cautious criterion as both necessary and sufficient, and even if we could then defend the rights of the sleeping or temporarily comatose (through the use of counterfactuals, for example), how could we possibly save the permanently comatose or the severely retarded? What rights could these 'human vegetables' possess? Since species is not a morally relevant factor, they obviously possess the same rights as nonhuman vegetables: none.

It is thus evident that from the perspective of personhood strict moral reasoning must condone at least feticide and infanticide, and potentially other killings as well, pending only the establishment of the requisite sufficient conditions.

Let us now examine the implications of personhood from another angle. Tooley maintains that if A has a right to X, then A is the sort of thing that is a subject of experiences and other mental states. A is capable of desiring X, and it is wrong for others to deprive A of X if A desires it. Tooley concludes from this that feticide and infanticide are permissible (since fetuses and infants, who do not possess a concept of continuing self, cannot desire to live), but that killing adult animals (non-*Homo sapiens*), is not permissible (if they possess a concept of continuing self). He also believes that whereas it is permissible to kill a kitten (since it cannot desire to live), it is not permissible to torture it (since it can desire not to suffer pain). On his view, this means that though a kitten has no right to life, it does have a right not to be tortured. In other words, since it is wrong for others to torture the kitten given that it desires not to be tortured, the kitten must have the *right* not to be tortured. [15]

This is the sort of right—the right that derives from someone else's wrong—that Thomson discusses at length. According to Thomson, we use the word 'wrong' to cover a variety of activities, not all of which involve a violation of rights. There are acts that may indeed be 'heinously evil' (as Tooley characterizes the torturing of a kitten), but which nevertheless do not imply infringement of rights. And there are degrees of wrong. There is the wrong of the Good Samaritan, the wrong of the Minimally Decent Samaritan, and the wrong of one who violates another's rights. We are thus not justified in deducing rights from what is wrong. Is it in fact *only* wrong to violate rights?

On Tooley's view, or any view in which rights form the sole basis of morality, this is indeed the case. It must turn out that unless kittens have the right not to be tortured there is nothing wrong with torturing them, and unless fetuses, infants, kittens, the permanently comatose, the severely retarded (and others?) have the right to life, it is not wrong to kill them at will.

The Importance of Duties

If these are indeed the implications of a personhood morality, perhaps such a morality is inadequate. Let us consider the following situations described by Wertheimer: "I stumble in the dark over my sleeping schnauzer; I stumble over my ottoman: To *blame* either nonperson is irrational; to blame the dog is also *unfair*, but to blame the funiture is neither fair nor *unfair*. So too: my bitch leaves me five pups. Without special reason it would be unfair to apportion the food unequally among them."[16] Would we wish to say in the first instance that a dog has a *right* not to be blamed, or, in the second, that the pups have a *right* to an equal share of the food?

Two possible answers to this question suggest themselves. The first, which accords with Tooley's intuition, is that if the dog or the pups have no such rights, then it is a mistake to say that it is 'unfair' to blame the dog or 'unfair' to distribute the food unequally among the pups. The second, which accords with my intuition, is that we can consistently affirm the unfairness of these activities while at the same time denying the dog and pups any *rights* to have us refrain from such activities.

But there is more here than a difference of intuitions. Hart, in his article "Are There Any Natural Rights?"[17] points to a confusion in the view that for every wrong there is a corresponding right that has been violated. If X promises Y to look after Y's mother, then it is surely wrong for X not to look after Y's mother. But there is more than one wrong here: X has wronged both Y and Y's mother; he has broken a promise to Y and has neglected Y's mother (who, I suppose, needs looking after and stands to suffer by not being looked after). Clearly, only one of these wrongs involves the violation of a right, and that is the broken promise. The other does not, since Y's mother has no rights with respect to X. We may thus distinguish two kinds of wrongs: (1) the wrong that involves violation of a right, and (2) the wrong that adversely affects a victim.

As moral agents, we must be concerned about every activity that brings harm. We have duties *with respect to* beings even if they have no rights against us. Such duties are not difficult to justify. Becker points out that our duties may derive from our special roles as parents, teachers, friends, doctors, etc., or from the consequences for the agent or society. They also "may be justified as a requirement of those patterns of life or character traits of which we can justifiably approve, morally."[18] In none of these justifications is there any need to appeal to rights.

In terms of the abortion issue, a shift of emphasis from the rights of fetuses to the duties of moral agents must have a tremendous impact. If we can have duties concerning fetuses even if they are not persons, abortion may turn out to be absolutely impermissible. And this surely holds true for infanticide as well.

There is one argument in support of abortion that does present a problem even for an agent-centered morality: Since (1) fetuses are potential persons, (2) potential persons do not exist, and (3) entities that do not exist cannot be the objects of duties, then (4) fetuses cannot be the objects of duties.[19]

Even if we grant the truth of premise (1), which is acceptable to Warren and might be acceptable to Tooley as well, and premise (2), which surely requires much elucidation, the validity of the argument remains doubtful because of an ambiguity in premise (3). The notion of being an 'object of duties' contained in premise (3) could mean one of two things. It could mean either that since one has duties only to things that have rights, these rights-possessing things must exist in order to be objects of duties, or it could mean that one cannot have duties even *with respect to* things that do not exist. It seems to me that whereas the former is true, the latter surely is not. Rights do *belong to* individuals, and unless those individuals exist, the duties corresponding to these rights cannot arise. However, as long as we recognize a sense in which we can have duties without recourse to corresponding rights, we need not locate existing individuals to whom these rights belong. Thus, for example, we may have duties with respect to future generations (e.g., the duty not to pollute water and air); future generations do not exist and hence do not have rights, but nevertheless they do qualify, in one sense, as objects of duties.

There yet remain two serious objections to the idea that a morality in which all duties have correlative rights is inadequate. Both are suggested by William Frankena in response to the Hart essay mentioned above. If, says Frankena, these so-called duties which have no corresponding rights are introduced because, unlike ordinary duties (which appeal, according to Hart, to the equal right to freedom), they appeal to actual or possible benefits or suffering, then this is nothing but the old distinction between justice and benevolence. And if they are introduced because they, as opposed to ordinary duties, are not legislatable, it must be pointed out that there are in fact ordinary duties in which legal coercion would not be appropriate. [20]

However, Frankena is surely mistaken if he supposes that duties which have no corresponding rights but which *do have victims* are to be considered mere benevolence. *Justice* may demand that we refrain from harming or mistreating even those who have no claims on us. It is not benevolence that requires that we not torture kittens or that we distribute food equally among pups or that we not blame the schnauzer over which we stumble.

Also, far from being of no concern to the law, these duties have been and should be included in the law. Thomson, for example, believes that Minimally Decent Samaritanism has been legislated elsewhere and ought to be legislated here,[21] and Hart points out that important legal codes, such as the Decalog, make no reference whatsoever to rights. In fact, he says: "Most natural law thinkers down to

Hooker conceived of natural law in this way: there were natural duties compliance with which would certainly benefit man—things to be done to achieve man's natural end—but not natural rights."[22]

If it is indeed possible to legislate duties with no corresponding rights, issues that were closed under personhood once again become vital questions. Intriguing questions such as these raised by Thomson—Should we distinguish between a pregnancy due to rape and a pregnancy for which the mother is responsible? Would it make any difference if a woman were pregnant for nine years instead of nine months? Does the mother's physical or emotional illness constitute a justification for abortion?—are dead ends in terms of rights.[23] Even the case in which the fetus threatens its mother's life has no easy solution. The dubious principle of 'double-effect,' if operant, absolutely rules out abortion even to save the mother's life, and Thomson, in order to *justify* therapeutic abortion, appeals to a principle no less objectionable—the principle of self-defense.[24]

If, however, we approach the abortion issue from the perspective of duties rather than of rights, all these questions begin to allow and require reexamination. Are our responsibilities as moral agents greater with respect to a fetus that we intentionally create than with respect to one forced upon us by rape? Precisely how much hardship are we required to undergo for the sake of fetuses? Must we surrender our lives even though the fetus has no rights against us?

Similarly, in terms of duties, though not in terms of rights, the question "Is the fetus human?" is once more significant. For just as our duties to our families, friends, and countrymen are stronger than our duties to strangers, so might our duties to humans be stronger than our duties to nonhumans.

It is now also possible to make sense of the role the fetus's innocence might play in the abortion controversy. Right-to-lifers have often pointed to this innocence in demanding the prohibition of abortion. However, the innocence of the fetus does nothing to counterbalance the fact that it simply is not a person. But this same innocence may figure significantly in considerations regarding the duties of others toward it: it is likely that our duties toward noninnocents are not as great as those toward innocents.

Finally, as Thomson notes, "opponents of abortion have been so concerned to make out the independence of the fetus in order to establish that it has a right to life, just as its mother does, that they have tended to overlook the possible support they might gain from making out that the fetus is *dependent* on the mother."[25] Indeed, our responsibilitis to the fetus arise not out of its rights, but rather out of its needs and its total dependence on *us*.

Notes and References

[1]An especially enlightening article is Lawrence C. Becker's "Human Being: The Boundaries of the Concept." *Philosophy and Public Affairs* 4, no. 4 (Summer 1975), 334–358.

[2]*See* Roger Wertheimer, "Understanding the Abortion Argument," *Philosophy and Public Affairs* 1, no. 1 (Fall 1971), 67–95, esp. 70–75.

[3]Judith Jarvis Thomson, "A Defense of Abortion," *Philosophy and Public Affairs* 1, no. 1 (Fall 1971), 47–66.

[4]Mary Anne Warren, "On the Moral and Legl Status of Abortion," *The Monist* 57, no. 1 (January 1973), 43–61; Michael Tooley, "Abortion and Infanticide," *Philosopy and Public Affairs* 2, no. 1 (Fall 1972), 37–65.

[5]*See*, for example, works on abortion by such philosophers as J. Thomson, R. Wertheimer, R. M. Hare, G. Grisez, and B. Brody, to name just a few.

[6]Becker, pp. 334–335.

[7]Becker incorrectly assumes that personhood is such a contribution and hence makes the following unfair accusation: "It taxes the concept of membership in the species too far to say that a 14-year-old, so catastrophically deficient as to warrant the claim that he or she is not a 'person,' is not a member of the species." (p. 348).

[8]This term is borrowed from Peter Singer, "Animal Liberation," review of Stanley and Roslind Godlovitch and John Harris, eds., *Animals, Men and Morals*, in *New York Review of Books* (April 5, 1973), 17–21.

[9]R. M. Hare, "Abortion and the Golden Rule." *Philosophy and Public Affairs* 4, no. 3 (Spring 1975), 203. Since Hare believes that our intuitions result from our upbringing, he urges that we resist the temptation to allow our ethical positions to be based on an appeal to intuition.

[10]Tooley, p. 44.

[11]Had their objective not been to put abortion on a par morally with getting a haircut, Warren and Tooley might have been content with Thomson's argument (which they both applaud) that granting the fetus a right to life does not automatically render all abortion impermissible. But, as Warren explicitly states, the flaw in Thomson's view is that, in treating abortion as a morally complex issue, it remains unacceptable to opponents of restrictive abortion laws for whom abortion is more like a morally neutral act (*see* Warren, p. 52 and Tooley, p. 52).

[12]Infants at least fulfill the first and third of Warren's criteria.

[13]Warren added a "Postscript on Infanticide" to a reprint of her article which appears in Richard Wasserstrom's anthology *Today's Moral Problems* (New York: Macmillan Co.; 1975), pp. 120–136. In it she states unequivocally that an infant has no right to life, but she does offer two arguments in opposition to in-

fanticide. The first is that at least in this country and in this period of history there are people who wish to have newborn babies unwanted by their parents. "Thus, infanticide is wrong for reasons analogous to those which make it wrong to wantonly destroy natural resources or great works of art" (p.135). Second, as long as there are people who are willing to pay additional taxes in order to provide the means to care for unwanted infants, it is wrong to destroy them. (These arguments do not apply to feticide because fetuses violate their mothers' rights to freedom, happiness, and self-determination, and the rights of these mothers override the rights of those who would like the fetuses preserved.) Of course, however, since the infant has no right to life, should an occasion arise in which society cannot afford to, or is unwilling to, care for an unwanted or defective infant, its destruction would be permissible. Technically, Tooley does not regard the term 'person' as synonymous with 'one who has rights': he restricts it so that it applies only to those who have a 'serious right to life.' The reason he gives for this is that not everything that has rights necessarily has a serious right to life. But that does not explain why he cannot have the term 'person' refer to anything that has rights and some *other* term refer only to things that have a serious right to life. In fact, his criterion for 'personhood' is just his general criterion for having rights applied to the specific right to life, as follows: To have a right to X, one must be capable of desiring X; to have a right to life, therefore, one must be capable of desiring to live.

[14]This is so common a criterion for having rights that Becker fails to notice that the argument for personhood need not rest on it. (Tooley makes a point of avoiding it.) Since Becker (p. 349) lodges his attack against this criterion, his criticisms of 'personhood' are largely ineffectual.

[15]Tooley thinks that the notion that a kitten's rights are determined (in part) by what it can desire helps to account for the phenomenon "that while in the case of adult humans most people would consider it worse to kill an individual than to torture him for an hour, we do not usually view the killing of a newborn kitten as morally outrageous, although we would regard someone who tortured a newborn kitten for an hour as heinouly evil" (p. 63). It is no accident that Tooley chooses to contrast an adult human with a *newborn kitten*, for were we to substitute 'adult cat' for 'newborn kitten,' our moral intuition would most likely remain unchanged, and yet, on Tooley's view, it should change: Adult cats, unlike newborn kittens, do fulfill the necessary criterion for having a serious right to life.

[16]Wertheimer, p. 90, n.

[17]Hart, "Are There Any Natural Rights?" *Philosophical Review* 64, no. 2 (April 1955), 175–191.

[18]Becker, p. 350.

[19]This argument is cited by Hare in "Abortion and the Golden Rule" (p. 219), but is not supported by him.

[20]*See* William K. Frankena's "Natural and Inalienable Rights," *Philosophical Review* 64, no. 2 (April 1955), 215.

[21]Thomson, p. 63.

[22]Hart, p. 182.

[23]Thomson, in order to answer these questions in terms of rights, is forced to define 'right to life' as the 'right not be killed *unjustly.* ' In this definition, 'unjustly'

does all the work Thomson assigns to it in order to secure her view that abortion is permissible in the case of rape for whatever reason and in no other case except when the life of the mother is threatened. Thus is is not 'unjust' to kill a fetus that is a product of rape or one that treatens its mother's life, and the fetus's right to life remains intact.

[24]Baruch Brody in his article "Thomson on Abortion," *Philosophy and Public Affairs* **1**, no. 3 (Spring 1972), 335–340, disputes Thomson's claim that, in self-defense, a woman may procure an abortion when the fetus threatens her life. According to Brody, in order for X to be justified in killing Y in self-defense, at least two conditions must obtain: (1) The continued existence of Y must pose a threat to X's life, a threat that can be met only by the taking of Y's life. (2) Y must be unjustly attempting to take X's life. Since in the abortion case only the first of these conditions is met, self-defense remains unjustified. The only time it is permissible to take Y's life when only the first condition is satisfied, according to Brody, is if Y is going to die shortly anyway.

[25]Thomson, p. 58.

Freedom of the Will
and the Concept of a Person

Harry G. Frankfurt

What philosophers have lately come to accept as analysis of the concept of a person is not actually analysis of that concept at all. Strawson, whose usage represents the current standard, identifies the concept of a person as "the concept of a type of entity such that both predicates ascribing states of consciousness and predicates ascribing corporeal characteristics...are equally applicable to a single individual of that single type."[1] But there are many entities besides persons that have both mental and physical properties. As it happens—though it seems extraordinary that this should be so—there is no common English word for the type of entity Strawson has in mind, a type that includes not only human beings but animals of various lesser species as well. Still, this hardly justifies the misappropriation of a valuable philosophical term.

Whether the members of some animal species are persons is surely not to be settled merely by determining whether it is correct to apply to them, in addition to predicates ascribing corporeal characteristics, predicates that ascribe states of consciousness. It does violence to our language to endorse the application of the term 'person' to those numerous creatures which do have both psychological and material properties, but which are manifestly not persons in any normal sense of the word. This misuse of language is doubtless innocent of any theoretical error. But although the offense is "merely verbal," it does significant harm. For it gratuitously diminishes our philosophical vocabulary, and it increases the likelihood that we will overlook the important area of inquiry with which the term 'person' is most naturally associated. It might have been expected that no problem would be of more central and per-

127

sistent concern to philosophers than that of understanding what we ourselves essentially are. Yet this problem is so generally neglected that it has been possible to make off with its very name almost without being noticed and, evidently, without evoking any widespread feeling of loss.

There is a sense in which the word 'person' is merely the singular form of 'people' and in which both terms connote no more than membership in a certain biological species. In those senses of the word which are of greater philosophical interest, however, the criteria for being a person do not serve primarily to distinguish the members of our own species from the members of other species. Rather, they are designed to capture those attributes that are the subject of our most humane concern with ourselves and the source of what we regard as most important and most problematical in our lives. Now these attributes would be of equal significance to us even if they were not in fact peculiar and common to the members of our own species. What interests us most in the human condition would not interest us less if it were also a feature of the condition of other creatures as well.

Our concept of ourselves as persons is not to be understood, therefore, as a concept of attributes that are necessarily species-specific. It is conceptually possible that members of novel or even of familiar nonhuman species should be persons; and it is also conceptually possible that some members of the human species are not persons. We do in fact assume, on the other hand, that no member of another species is a person. Accordingly, there is a presumption that what is essential to persons is a set of characteristics that we generally suppose—whether rightly or wrongly—to be uniquely human.

It is my view that one essential difference between persons and other creatures is to be found in the structure of a person's will. Human beings are not alone in having desires and motives, or in making choices. They share these things with the members of certain other species, some of whom even appear to engage in deliberation and to make decisions based upon prior thought. It seems to be peculiarly characteristic of humans, however, that they are able to form what I shall call "second-order desires" or "desires of the second order."

Besides wanting and choosing and being moved *to do* this or that, men may also want to have (or not to have) certain desires and motives. They are capable of wanting to be different, in their preferences and purposes, from what they are. Many animals appear to have the capacity for what I shall call "first-order desires" or "desires of the first order," which are simply desires to do or not to do one thing or another. No animal other than man, however, appears to have the capacity for reflective self-evaluation that is manifested in the formation of second-order desires.[2]

I

The concept designated by the verb 'to want' is extraordinarily elusive. A statement of the form "*A* wants to *X*"—taken by itself, apart from a context that serves to amplify or to specify its meaning—conveys remarkably little information. Such a statement may be consistent, for example, with each of the following statements: (a) the prospect of doing *X* elicits no sensation or introspectible emotional response in *A;* (b) *A* is unaware that he wants to *X;* (c) *A* believes that he does not want to *X;* (d) *A* wants to refrain from *X*-ing; (e) *A* wants to *Y* and believes that it is impossible for him both to *Y* and to *X;* (f) *A* does not "really" want to *X;* (g) *A* would rather die than *X;* and so on. It is therefore hardly sufficient to formulate the distinction between first-order and second-order desires, as I have done, by suggesting merely that someone has a first-order desire when he wants to do or not to do such-and-such, and that he has a second-order desire when he wants to have or not to have a certain desire of the first order.

As I shall understand them, statements of the form "*A* wants to *X*" cover a rather broad range of possibilities.[3] They may be true even when statements like (a) through (g) are true: when *A* is unaware of any feelings concerning *X*-ing, when he is unaware that he wants to *X,* when he deceives himself about what he wants and believes falsely that he does not want to *X,* when he also has other desires that conflict with his desire to *X,* or when he is ambivalent. The desires in question may be conscious or unconscious, they need not be univocal, and *A* may be mistaken about them. There is a further source of uncertainty with regard to statements that identify someone's desires, however, and here it is important for my purposes to be less permissive.

Consider first those statements of the form "*A* wants to *X*" which identify first-order desires—that is, statements in which the term 'to

X' refers to an action. A statement of this kind does not, by itself, indicate the relative strength of A's desire to X. It does not make it clear whether this desire is at all likely to play a decisive role in what A actually does or tries to do. For it may correctly be said that A wants to X even when his desire to X is only one among his desires and when it is far from being paramount among them. Thus, it may be true that A wants to X when he strongly prefers to do something else instead; and it may be true that he wants to X despite the fact that, when he acts, it is not the desire to X that motivates him to do what he does. On the other hand, someone who states that A wants to X may mean to convey that it is this desire that is motivating or moving A to do what he is actually doing or that A will in fact be moved by this desire (unless he changes his mind) when he acts.

It is only when it is used in the second of these ways that, given the special usage of 'will' that I propose to adopt, the statement identifies A's will. To identify an agent's will is either to identify the desire (or desires) by which he is motivated in some action he performs or to identify the desire (or desires) by which he will or would be motivated when or if he acts. An agent's will, then, is identical with one or more of his first-order desires. But the notion of the will, as I am employing it, is not coextensive with the notion of first-order desires. It is not the notion of something that merely inclines an agent in some degree to act in a certain way. Rather, it is the notion of an *effective* desire—one that moves (or will or would move) a person all the way to action. Thus the notion of the will is not coextensive with the notion of what an agent intends to do. For even though someone may have a settled intention to do X, he may nonetheless do something else instead of doing X because, despite his intention, his desire to do X proves to be weaker or less effective than some conflicting desire.

Now consider those statements of the form "A wants to X" which identify second-order desires—that is, statements in which the term 'to X' refers to a desire of the first order. There are also two kinds of situation in which it may be true that A wants to want to X. In the first place, it might be true of A that he wants to have a desire to X despite the fact that he has a univocal desire, altogether free of conflict and ambivalence, to refrain from X-ing. Someone might want to have a certain desire, in other words, but univocally want that desire to be unsatisfied.

Suppose that a physician engaged in psychotherapy with narcotics addicts believes that his ability to help his patients would be enhanced if he understood better what it is like for them to desire the drug to which they are addicted. Suppose that he is led in this way to want to have a desire for the drug. If it is a genuine desire that he wants, then what he wants is not merely to feel the sensations that addicts characteristically feel when they are gripped by their desires for the drug. What the physician wants, insofar as he wants to have a desire, is to be inclined or moved to some extent to take the drug.

It is entirely possible, however, that, although he wants to be moved by a desire to take the drug, he does not want this desire to be effective. He may not want it to move him all the way to action. He need not be interested in finding out what it is like to take the drug. And insofar as he now wants only to *want* to take it, and not to *take* it, there is nothing in what he now wants that would be satisfied by the drug itself. He may now have, in fact, an altogether univocal desire *not* to take the drug; and he may prudently arrange to make it impossible for him to satisfy the desire he would have if his desire to want the drug should in time be satisfied.

It would thus be incorrect to infer, from the fact that the physician now wants to desire to take the drug, that he already does desire to take it. His second-order desire to be moved to take the drug does not entail that he has a first-order desire to take it. If the drug were now to be administered to him, this might satisfy no desire that is implicit in his desire to want to take it. While he wants to want to take the drug, he may have *no* desire to take it; it may be that *all* he wants is to taste the desire for it. That is, his desire to have a certain desire that he does not have may not be a desire that his will should be at all different than it is.

Someone who wants only in this truncated way to want to X stands at the margin of preciosity, and the fact that he wants to want to X is not pertinent to the identification of his will. There is, however, a second kind of situation that may be described by '*A* wants to want to X; and when the statement is used to describe a situation of this kind, then it does not pertain to what *A* wants his will to be. In such cases, the statement means that *A* wants the desire to X to be the desire that moves him effectively to act. It is not merely that he wants the desire to X to be among the desires by which, to one degree or another, he is moved or inclined to act. He wants this desire to be effective—that is, to provide the motive in what he actually does. Now when the statement that *A* wants to *X* is used in this way, it does entail that *A* already has a desire to *X*. It could not be true both that *A* wants the desire to *X* to move him into action and that he does not want to *X*. It is only if he does want to *X* that he can coherently want the desire to *X* not merely to be one of his desires but, more decisively, to be his will.[4]

Suppose a man wants to be motivated in what he does by the desire to concentrate on his work. It is necessarily true, if this supposition is correct, that he already wants to concentrate on his work. This desire is now among his desires. But the question of whether or not his second-order desire is fulfilled does not turn merely on whether the desire he wants is one of his desires. It turns on whether this desire is, as he wants it to be, his effective desire or will. If, when the chips are down, it is his desire to concentrate on his work that moves him to do what he does, then what he wants at that time is indeed (in the relevant sense) what he wants to want. If it is some other desire that actually moves him when he acts, on the other hand, then what he wants at that time is not (in the relevant sense) what he wants to want. This will be so despite the fact that the desire to concentrate on his work continues to be among his desires.

II

Someone has a desire of the second order either when he wants simply to have a certain desire or when he wants a certain desire to be his will. In situations of the latter kind, I shall call his second-order desires "second-order volitions" or "volitions of the second order." Now it is having second-order volitions, and not having second-order desires generally, that I regard as essential to being a person. It is logically possible, however unlikely, that there should be an agent with second-order desires but with no volitions of the second order. Such a creature, in my view, would not be a person. I shall use the term "wanton" to refer to agents who have first-order desires but who are not persons because, whether or not they have desires of the second order, they have no second-order volitions.[5]

The essential characteristic of a wanton is that he does not care about his will. His desires move him to do certain things, without its being true of him either that he wants to be moved by those desires or that he prefers be moved by other desires. The class of wantons includes all nonhuman animals that have desires and all very young children. Perhaps it also includes some adult human beings as well. In any case, adult humans may be more or less wanton; they may act wantonly, in response to first-order desires concerning which they have no volitions of the second order, more or less frequently.

The fact that a wanton has no second-order volitions does not mean that each of his first-order desires is translated heedlessly and at once into action. He may have no opportunity to act in accordance with some of his desires. Moreover, the translation of his desires into action may be delayed or precluded either by conflicting desires of the first order or by the intervention of deliberation. For a wanton may possess and employ rational faculties of a high order. Nothing in the concept of a wanton implies that he cannot reason or that he cannot deliberate concerning how to do what he wants to do. What distinguishes the rational wanton from other rational agents is that he is not concerned with the desirability of his desires themselves. He ignores the question of what his will is to be. Not only does he pursue whatever course of action he is most strongly inclined to pursue, but he does not care which of his inclinations is the strongest.

Thus a rational creature, who reflects upon the suitability to his desires of one course of action or another, may nonetheless be a wanton. In maintaining that the essence of being a person lies not in reason but in will, I am far from suggesting that a creature without reason may be a person. For it is only in virtue of his rational capacities that a person is capable of becoming critically aware of his own will and of forming volitions of the second order. The structure of a person's will presupposes, accordingly, that he is a rational being.

The distinction between a person and a wanton may be illustrated by the difference between two narcotics addicts. Let us suppose that the physiological condition accounting for the addiction is the same in both men, and that both succumb inevitably to their periodic desires for the drug to which they are addicted. One of the addicts hates his addiction and always struggles desperately, although to no avail, against its thrust. He tries everything that he thinks might enable him to overcome his desires for the drug. But these desires are too powerful for him to withstand and invariably, in the end, they conquer

him. He is an unwilling addict, helplessly violated by his own desires.

The unwilling addict has conflicting first-order desires: he wants
to take the drug, and he also wants to refrain from taking it. In addition to these first-order desires, however, he has a volition of the
second order. He is not a neutral with regard to the conflict between
his desire to take the drug and his desire to refrain from taking it. It
is the latter desire, and not the former, that he wants to constitute his
will; it is the latter desire, rather than the former, that he wants to be
effective and to provide the purpose that he will seek to realize in
what he actually does.

The other addict is a wanton. His actions reflect the economy of
his first-order desires, without his being concerned whether the desires that move him to act are desires by which he wants to be moved
to act. If he encounters problems in obtaining the drug or in admini-
stering it to himself, his responses to his urges to take it may involve
deliberation. But it never occurs to him to consider whether he
wants the relations among his desires to result in his having the will
he has. The wanton addict may be an animal, and thus incapable of
being concerned about his will. In any event, he is, in respect of his
wanton lack of concern, no different from an animal.

The second of these addicts may suffer a first-order conflict similar to the first-order conflict suffered by the first. Whether he is
human or not, the wanton may (perhaps due to conditioning) both
want to take the drug and want to refrain from taking it. Unlike the
unwilling addict, however, he does not prefer that one of his conflicting desires should be paramount over the other; he does not prefer that one first-order desire rather than the other should constitute
his will. It would be misleading to say that he is neutral as to the
conflict between his desires, since this would suggest that he regards
them as equally acceptable. Since he has no identity apart from his
first-order desires, it is true neither that he prefers one to the other
nor that he prefers not to take sides.

It makes a difference to the unwilling addict, who is a person,
which of his conflicting first-order desires wins out. Both desires
are his, to be sure; and whether he finally takes the drug or finally
succeeds in refraining from taking it, he acts to satisfy what is in a
literal sense his own desire. In either case he does something he
himself wants to do, and he does it not because of some external in-

fluence whose aim happens to coincide with his own but because of his desire to do it. The unwilling addict identifies himself, however, through the formation of a second-order volition, with one rather than with the other of his conflicting first-order desires. He makes one of them more truly his own and, in so doing, he withdraws himself from the other. It is in virtue of this identification and withdrawal, accomplished through the formation of a second-order volition, that the unwilling addict may meaningfully make the analytically puzzling statements that the force moving him to take the drug is a force other than his own, and that it is not of his own free will but rather against his will that this force moves him to take it.

The wanton addict cannot or does not care which of his conflicting first-order desires wins out. His lack of concern is not due to his inability to find a convincing basis for preference. It is due either to his lack of the capacity for reflection or to his mindless indifference to the enterprise of evaluating his own desires and motives.[6] There is only one issue in the struggle to which his first-order conflict may lead: whether the one or the other of his conflicting desires is the stronger. Since he is moved by both desires, he will not be altogether satisfied by what he does no matter which of them is effective. But it makes no difference *to him* whether his craving or his aversion gets the upper hand. He has no stake in the conflict between them and so, unlike the unwilling addict, he can neither win nor lose the struggle in which he is engaged. When a *person* acts, the desire by which he is moved is either the will he wants or a will he wants to be without. When a *wanton* acts, it is neither.

III

There is a very close relationship between the capacity for forming second-order volitions and another capacity that is essential to persons—one that has often been considered a distinguishing mark of the human condition. It is only because a person has volitions of the second order that he is capable both of enjoying and of lacking freedom of the will. The concept of a person is not only, then, the concept of a type of entity that has both first-order desires and volitions of the second order. It can also be construed as the concept of a type of entity for whom the freedom of its will may be a problem.

This concept excludes all wantons, both infrahuman and human, since they fail to satisfy an essential condition for the enjoyment of freedom of the will. And it excludes those suprahuman beings, if any, whose wills are necessarily free.

Just what kind of freedom is the freedom of the will? This question calls for an identification of the special area of human experience to which the concept of freedom of the will, as distinct from the concepts of other sorts of freedom, is particularly germane. In dealing with it, my aim will be primarily to locate the problem with which a person is most immediately concerned when he is concerned with the freedom of his will.

According to one familiar philosophical tradition, being free is fundamentally a matter of doing what one wants to do. Now the notion of an agent who does what he wants to do is by no means an altogether clear one; both the doing and the wanting, and the appropriate relation between them as well, require elucidation. But although its focus needs to be sharpened and its formulation refined, I believe that this notion does capture at least part of what is implicit in the idea of an agent who acts freely. It misses entirely, however, the peculiar content of the quite different idea of an agent whose will is free.

We do not suppose that animals enjoy freedom of the will, although we recognize that an animal may be free to run in whatever direction it wants. Thus, having the freedom to do what one wants to do is not a sufficient condition of having a free will. It is not a necessary condition either. For to deprive someone of his freedom of action is not necessarily to undermine the freedom of his will. When an agent is aware that there are certain things he is not free to do, this doubtless affects his desires and limits the range of choices he can make. But suppose that someone, without being aware of it, has in fact lost or been deprived of his freedom of action. Even though he is no longer free to do what he wants to do, his will may remain as free as it was before. Despite the fact that he is not free to translate his desires into actions or to act according to the determinations of his will, he may still form those desires and make those determinations as freely as if his freedom of action had not been impaired.

When we ask whether a person's will is free we are not asking whether he is in a position to translate his first-order desires into actions. That is the question of whether he is free to do as he pleases.

The question of the freedom of his will does not concern the relation between what he does and what he wants to do. Rather, it concerns his desires themselves. But what question about them is it?

It seems to me both natural and useful to construe the question of whether a person's will is free in close analogy to the question of whether an agent enjoys freedom of action. Now freedom of action is (roughly, at least) the freedom to do what one wants to do. Analogously, then, the statement that a person enjoys freedom of the will means (also roughly) that he is free to want what he wants to want. More precisely, it means that he is free to will what he wants to will, or to have the will he wants. Just as the question about the freedom of an agent's action has to do with whether it is the action he wants to perform, so the question about the freedom of his will has to do with whether it is the will he wants to have.

It is in securing the conformity of his will to his second-order volitions, then, that a person exercises freedom of the will. And it is in the discrepancy between his will and his second-order volitions, or in his awareness that their coincidence is not his own doing but only a happy chance, that a person who does not have this freedom feels its lack. The unwilling addict's will is not free. This is shown by the fact that it is not the will he wants. It is also true, though in a different way, that the will of the wanton addict is not free. The wanton addict neither has the will he wants nor has a will that differs from the will he wants. Since he has no volitions of the second order, the freedom of his will cannot be a problem for him. He lacks it, so to speak, by default.

People are generally far more complicated than my sketchy account of the structure of a person's will may suggest. There is as much opportunity for ambivalence, conflict, and self-deception with regard to desires of the second order, for example, as there is with regard to first-order desires. If there is an unresolved conflict among someone's second-order desires, then he is in danger of having no second-order volition; for unless this conflict is resolved, he has no preference concerning which of his first-order desires is to be his will. This condition, if it is so severe that it prevents him from identifying himself in a sufficiently decisive way with *any* of his conflicting first-order desires, destroys him as a person. For it either tends to paralyze his will and to keep him from acting at all, or it tends to remove him from his will so that his will operates without his participation. In both cases he becomes, like the unwilling addict though in a different way, a helpless bystander to the forces that move him.

Another complexity is that a person may have, especially if his second-order desires are in conflict, desires and volitions of a higher order than the second. There is no theoretical limit to the length of the series of desires of higher and higher orders; nothing except common sense and, perhaps, a saving fatigue prevents an individual from obsessively refusing to identify himself with any of his desires until he forms a desire of the next higher order. The tendency to generate such a series of acts of forming desires, which would be a case of humanization run wild, also leads toward the destruction of a person.

It is possible, however, to terminate such a series of acts without cutting it off arbitrarily. When a person identifies himself *decisively* with one of his first-order desires, this commitment "resounds" throughout the potentially endless array of higher orders. Consider a person who, without reservation or conflict, wants to be motivated by desire to concentrate on his work. The fact that his second-order volition to be moved by this desire is a decisive one means that there is no room for questions concerning the pertinence of desires or volitions of higher orders. Suppose the person is asked whether he wants to want to want to concentrate on his work. He can properly insist that this question concerning a third-order desire does not arise. It would be a mistake to claim that, because he has not considered whether he wants the second-order volition he has formed, he is indifferent to the question of whether it is with this volition or with some other that he wants his will to accord. The decisiveness of the commitment he has made means that he has decided that no further question about his second-order volition, at any higher order, remains to be asked. It is relatively unimportant whether we explain this by saying that this commitment implicitly generates an endless series of confirming desires of higher orders, or by saying that the commitment is tantamount to a dissolution of the pointedness of all questions concerning higher orders of desire.

Examples such as the one concerning the unwilling addict may suggest that volitions of the second order, or of higher orders, must be formed deliberately and that a person characteristically struggles to ensure that they are satisfied. But the conformity of a person's will to his higher-order volitions may be far more thoughtless and spontaneous than this. Some people are naturally moved by kindness when they want to be kind, and by nastiness when they want to be nasty, without any explicit forethought and without any need for energetic self-control. Others are moved by nastiness when they want to be kind and by kindness when they intend to be nasty, equally without forethought and without active resistance to these violations of their higher-order desires. The enjoyment of freedom comes easily to some. Others must struggle to achieve it.

IV

My theory concerning the freedom of the will accounts easily for our disinclination to allow that this freedom is enjoyed by the members of any species inferior to our own. It also satisfies another condition that must be met by any such theory, by making it apparent why the freedom of the will should be regarded as desirable. The enjoyment of a free will means the satisfaction of certain desires—desires of the second or of higher orders—whereas its absence means their frustration. The satisfactions at stake are those that accrue to a person of whom it may be said that his will is his own. The corresponding frustrations are those suffered by a person of whom it may be said that he is estranged from himself, or that he finds himself a helpless or a passive bystander to the forces that move him.

A person who is free to do what he wants to do may yet not be in a position to have the will he wants. Suppose, however, that he enjoys both freedom of action and freedom of the will. Then he is not only free to do what he wants to do; he is also free to want what he wants to want. It seems to me that he has, in that case, all the freedom it is possible to desire or to conceive. There are other good things in life, and he may not possess some of them. But there is nothing in the way of freedom that he lacks.

It is far from clear that certain other theories of the freedom of the will meet these elementary but essential conditions: that it be understandable why we desire this freedom and why we refuse to ascribe it to animals. Consider, for example, Roderick Chisholm's quaint version of the doctrine that human freedom entails an absence of causal determination.[7] Whenever a person performs a free action, according to Chisholm, it's a miracle. The motion of a person's hand, when a person moves it, is the outcome of a series of physical causes; but some event in this series, "and presumably one of those that took place within the brain, was caused by the agent and not by any other events". A free agent has, therefore, "a prerogative which some would attribute only to God: each of us, when we act, is a prime mover unmoved".

This account fails to provide any basis for doubting that animals of subhuman species enjoy the freedom it defines. Chisholm says nothing that makes it seem less likely that a rabbit performs a miracle when it moves its leg than that a man does so when he moves his hand. But why, in any case, should anyone *care* whether he can interrupt the natural order of causes in the way Chisholm describes? Chisholm offers no reason for believing that there is a discernible difference between the experience of a man who miraculously initiates a series of causes when he moves his hand and a man who moves his hand without any such breach of the normal causal sequence. There appears to be no concrete basis for preferring to be involved in the one state of affairs rather than in the other.[8]

It is generally supposed that, in addition to satisfying the two conditions I have mentioned, a satisfactory theory of the freedom of the will necessarily provides an analysis of one of the conditions of moral responsibility. The most common recent approach to the problem of understanding the freedom of the will has been, indeed, to inquire what is entailed by the assumption that someone is morally responsible for what he has done. In my view, however, the relation between moral responsibility and the freedom of the will has been very widely misunderstood. It is not true that a person is morally responsible for what he has done only if his will was free when he did it. He may be morally responsible for having done it even though his will was not free at all.

A person's will is free only if he is free to have the will he wants. This means that, with regard to any of his first-order desires, he is free either to make that desire his will or to make some other first-order desire his will instead. Whatever his will, then, the will of the person whose will is free could have been otherwise; he could have done otherwise than to constitute his will as he did. It is a vexed question just how 'he could have done otherwise' is to be understood in contexts such as this one. But although this question is important to the theory of freedom, it has no bearing on the theory of moral responsibility. For the assumption that a person is morally responsible for what he has done does not entail that the person was in a position to have whatever will he wanted.

This assumption does entail that the person did what he did freely, or that he did it of his own free will. It is a mistake, however, to believe that someone acts freely only when he is free to do whatever he wants or that he acts of his own free will only if his will is free. Suppose that a person has done what he wanted to do, that he did it because he wanted to do it, and that the will by which he was moved when he did it was his will because it was the will he wanted. Then he did it freely and of his own free will. Even supposing that he could have done otherwise, he would not have done otherwise; and even supposing that he could have had a different will, he would not have wanted his will to differ from what it was. Moreover, since the will that moved him when he acted was his will because he wanted it to be, he cannot claim that his will was forced upon him or that he was a passive bystander to its constitution. Under these conditions, it is quite irrelevant to the evaluation of his moral responsibility to inquire whether the alternatives that he opted against were actually available to him.[9]

In illustration, consider a third kind of addict. Suppose that his addiction has the same physiological basis and the same irresistible thrust as the addictions of the unwilling and wanton addicts, but that he is altogether delighted with his condition. He is a willing addict, who would not have things any other way. If the grip of his addiction should somehow weaken, he would do whatever he could to reinstate it; if his desire for the drug should begin to fade, he would take steps to renew its intensity.

The willing addict's will is not free, for his desire to take the drug will be effective regardless of whether or not he wants this desire to constitute his will. But when he takes the drug, he takes it freely and of his own free will. I am inclined to understand his situation as involving the overdetermination of his first-order desire to take the drug. This desire is his effective desire because he is physiologically addicted. But it is his effective desire also because he wants it to be. His will is outside his control, but by his second-order desire that his desire for the drug should be effective, he has made this will his own.

Given that it is therefore not only because of his addiction that his desire for the drug is effective, he may be morally responsible for taking the drug.

My conception of the freedom of the will appears to be neutral with regard to the problem of determinism. It seems conceivable that it should be causally determined that a person is free to want what he wants to want. If this is conceivable, then it might be causally determined that a person enjoys a free will. There is no more than an innocuous appearance of paradox in the proposition that it is determined, ineluctably and by forces beyond their control, that certain people have free wills and that others do not. There is no incoherence in the proposition that some agency other than a person's own is responsible (even *morally* responsible) for the fact that he enjoys or fails to enjoy freedom of the will. It is possible that a person should be morally responsible for what he does of his own free will and that some other person should also be morally responsible for his having done it.[10]

On the other hand, it seems conceivable that it should come about by chance that a person is free to have the will he wants. If this is conceivable, then it might be a matter of chance that certain people enjoy freedom of the will and that certain others do not. Perhaps it is also conceivable, as a number of philosophers believe, for states of affairs to come about in a way other than by chance or as the outcome of a sequence of natural causes. If it is indeed conceivable for the relevant states of affairs to come about in some third way, then it is also possible that a person should in that third way come to enjoy the freedom of the will.

Notes

[1]Peter F. Strawson, *Individuals* (London), Methuen, 1959, pp. 101–102. Ayer's usage of 'person' is similar: "it is characteristic of persons in this sense that besides having various physical properties...they are also credited with various forms of consciousness" (A. J. Ayer, *The Concept of a Person*, New York: St. Martin's, 1963, p. 82). What concerns Strawson and Ayer is the problem of understanding the relation between mind and body, rather than the quite different problem of understanding what it is to be a creature that not only has a mind and a body but is also a person.

[2]For the sake of simplicity, I shall deal only with what someone wants or desires, neglecting related phenomena such as choices and decisions. I propose to use the verbs 'to want' and to 'desire' interchangeably, although they are by no

means perfect synonyms. My motive in forsaking the established nuances of these words arises from the fact that the verb 'to want,' which suits my purposes better so far as its meaning is concerned, does not lend itself so readily to the formation of nouns as does the verb 'to desire.' It is perhaps acceptable, albeit graceless, to speak in the plural of someone's "wants." But to speak in the singular of someone's "want" would be an abomination.

[3]What I say in this paragraph applies not only to cases in which 'to X ' refers to a possible action or inaction. It also applies to cases in which 'to X' refers to a first-order desire and in which the statement that 'A wants to X' is therefore a shortened version of a statement—"A wants to want to X'—that identifies a desire of the second order.

[4]It is not so clear that the entailment relation described here holds in certain kinds of cases, which I think may fairly be regarded as nonstandard, where the essential difference between the standard and the nonstandard cases lies in the kind of description by which the first-order desire in question is identified. Thus, suppose that A admires B so fulsomely that, even though he does not know what B wants to do, he wants to be effectively moved by whatever desire effectively moves B; without knowing what B's will is, in other words, A wants his own will to be the same. It certainly does not follow that A already has, among his desires, a desire like the one that constitutes B's will. I shall not pursue here the questions of whether there are genuine counterexamples to the claim made in the text or of how, if there are, that claim should be altered.

[5]Creatures with second-order desires but no second-order volitions differ significantly from brute animals, and, for some purposes, it would be desirable to regard them as persons. My usage, which withholds the designation 'person' from them, is thus somewhat arbitrary. I adopt it largely because it facilitates the formulation of some of the points I wish to make. Hereafter, whenever I consider statements of the form "A wants to want to X," I shall have in mind statements identifying second-order volitions and not statements identifying second-order desires that are not second-order volitions.

[6]In speaking of the evalution of his own desires and motives as being characteristic of a person, I do not mean to suggest that a person's second-order volitions necessarily manifest a *moral* stance on his part toward his first-order desires. It may not be from the point of view of morality that the person evaluates his first-order desires. Moreover, a person may be capricious and irresponsible in forming his second-order volitions and give no serious consideration to what is at stake. Second-order volitions express evaluations only in the sense that they are preferences. There is no essential restriction on the kind of basis, if any, upon which they are formed.

[7]"Freedom and Action," in K. Lehrer, ed., *Freedom and Determinism* (New York: Random House, 1966), pp. 11–44.

[8]I am not suggesting that the alleged difference between these two states of affairs is unverifiable. On the contrary, physiologists might well be able to show that Chisholm's conditions for a free action are not satisfied, by establishing that there is no relevant brain event for which a sufficient physical cause cannot be found.

⁹For another discussion of the considerations that cast doubt on the principle that a person is morally responsible for what he has done only if he could have done otherwise, see my "Alternate Possibilities and Moral Responsibility," *Journal of Philosophy* **LXVI**, 23 (Dec. 4, 1969), 829–839.

¹⁰There is a difference between being *fully* responsible and being *solely* responsible. Suppose that the willing addict has been made an addict by the deliberate and calculated work of another. Then it may be that both the addict and this other person are fully responsible for the addict's taking the drug, while neither of them is solely responsible for it. That there is a distinction between full moral responsibility and sole moral responsibility is apparent in the following example. A certain light can be turned on or off by flicking either of two switches, and each of these switches is simultaneously flicked to the "on" position by a different person, neither of whom is aware of the other. Neither person is solely responsible for the light's going on, nor do they share the responsibility in the sense that each is partially responsible; rather, each of them is fully responsible.

Conditions
of Personhood

Daniel Dennett

I am a person, and so are you. That much is beyond doubt. I am a human being, and *probably* you are too. If you take offense at the "probably" you stand accused of a sort of racism, for what is important about us is not that we are of the same biological species, but that we are both persons, and I have not cast doubt on that. One's dignity does not depend on one's parentage even to the extent of having been born of women or born at all. We normally ignore this and treat humanity as the deciding mark of personhood, no doubt because the terms are locally coextensive or almost coextensive. At this time and place human beings are the only persons we recognize, and we recognize almost all human beings as persons, but on the one hand we can easily contemplate the existence of biologically very different persons—inhabiting other planets, perhaps—and on the other hand we recognize conditions that exempt human beings from personhood, or at least some very important elements of personhood. For instance, infant human beings, mentally defective human beings, and human beings declared insane by licensed psychiatrists are denied personhood, or at any rate crucial elements of personhood.

One might well hope that such an important concept, applied and denied so confidently, would have clearly formulatable necessary and sufficient conditions for ascription, but if it does, we have not yet discovered them. In the end there may be none to discover. In the end we may come to realize that the concept of person is incoherent and obsolete. Skinner, for one, has suggested this, but the doctrine has not caught on, no doubt in part because it is difficult or even impossible to conceive of what it would be like if we abandoned the concept of a person. The idea that we might cease to view others and *ourselves* as persons (if it does not mean merely that we might annihilate ourselves, and hence cease to view anything as anything) is arguably self-contradictory.[1] So quite aside from whatever might be right or wrong in Skinner's grounds for his claim, it is hard to see how it could win out in contest with such an intuitively invulnerable notion. If then the concept of a person is in some way an ineliminable part of our conceptual scheme, it might still be in rather worse shape than we would like. It might turn out, for instance, that the concept of a person is only a free-floating honorific that we are all happy to apply to ourselves, and to others as the spirit moves us, guided by our emotions, aesthetic sensibilities, considerations of policy, and the like—just as those who are *chic* are all and only those who can get themselves considered *chic* by others who consider themselves *chic:* Being a person is certainly *something* like that, and if it were no more, we would have to reconsider, if we could, the importance with which we now endow the concept.

Supposing there *is* something more to being a person, the searcher for necessary and sufficient conditions may still have difficulties if there is more than one concept of a person, and there are grounds for suspecting this. Roughly, there seem to be two notions interwined here, which we may call the moral notion and the metaphysical notion. Locke says that "person"

> ...is a forensic term, appropriating actions and their merit; and so belongs only to intelligent agents, capable of a law, and happiness, and misery. This personality extends itself beyond present existence to what is past, only by consciousness—whereby it becomes concerned and accountable (*Essays,* Book II, Chap. XXVII).

Does the metaphysical notion—roughly, the notion of an intelligent, conscious, feeling agent—*coincide* with the moral notion—roughly, the notion of an agent who is accountable, who has both rights and responsibilities?   Or is it merely that being a person in the metaphysical sense is a necessary but not sufficient condition of being a person in the moral sense? Is being an entity to which states of consciousness or self-consciousness are ascribed *the same* as being an end-in-oneself, or is it merely one precondition? In Rawls's theory of justice, should the derivation from the original position be viewed as a demonstration of how metaphysical persons *can become* moral persons, or should it be viewed as a demonstration of why metaphysical persons *must be* moral persons?[2] In less technical surroundings the distinction stands out as clearly: when we declare a man insane we cease treating him as accountable, and we deny him most rights, but still our interactions with him are virtually indistinguishable from normal personal interactions unless he is very far gone in madness indeed. In one sense of "person," it seems, we continue to treat and view him as a person. I claimed at the outset that it was indubitable that you and I are persons. I could not plausibly hope—let alone aver—that all readers of this essay will be legally sane and morally accountable. What—if anything—was beyond all doubt may only have been that anything properly addressed by the opening sentence's personal pronouns, "you" and "I," was a person in the metaphysical sense. If that was all that was beyond doubt, then the metaphysical notion and the moral notion must be distinct. Still, even if we suppose there are these distinct notions, there seems every reason to believe that metaphysical personhood is a necessary condition of moral personhood.[3]

What I wish to do now is consider six familiar themes, each a claim to identify a necessary condition of personhood, and each, I think, a correct claim on some interpretation. What will be at issue here is first, how (on my interpretation) they are dependent on each other; second, why they are necessary conditions of moral person-hood, and third, why it is so hard to say whether they are jointly sufficient conditions for moral personhood. The *first* and most obvious theme is that persons are rational beings. It figures, for example, in the ethical theories of Kant and Rawls, and in the "metaphysical" theories of Aristotle and Hintikka.[4] The *second* theme is that persons are beings to which states of consciousness are attributed, or to which psychological or mental or *intentional predicates,* are ascribed. Thus Strawson identifies the concept of a person as "the concept of a type of entity such that *both* predicates ascribing states of consciousness *and* predicates ascribing corporeal characteristics" are applicable.[5] The *third* theme is that whether something counts as a person depends in some way on an *attitude taken* toward it, a *stance adopted* with respect to it. This theme suggests that it is not the case that once we have established the objective fact that something is a person we treat him or her or it a certain way, but that our treating him or her or it in this certain way is somehow and to some extent constitutive of its being a person. Variations on this theme have been stated by MacKay, Strawson, Amelie Rorty, Putnam, Sellars, Flew, Nagel, Van de Vate, and myself.[6] The *fourth* theme is that the object toward which this personal stance is taken must be capable of *reciprocating* in some way. Very different versions of this are expressed or hinted at by Rawls, MacKay, Strawson, Grice, and others. This reciprocity has sometimes been rather uninformatively expressed by the slogan: to be a person is to treat others as persons, and with this expression has often gone the claim that treating another as a person is treating him morally—perhaps obeying the Golden Rule, but this conflates different sorts of reciprocity. As Nagel says, "extremely hostile behavior toward another is compatible with treating him as a person" (p. 134), and as Van de Vate observes, one of the differences between some forms of manslaughter and murder is that the murderer treats the victim as a person.

The *fifth* theme is that persons must be capable of *verbal communication.* This condition handily excuses nonhuman animals from full personhood and the attendant moral responsibility, and seems at least implicit in all social contract theories of ethics. It is also a theme that has been stressed or presupposed by many writers in philosophy of mind, including myself, where the moral dimension of personhood has not been at issue. The *sixth* theme is that persons are distinguishable from other entities by being *conscious* in some special way: there is a way in which *we* are conscious in which no other species is conscious. Sometimes this is identified as *self*-consciousness of one sort or another. Three philosophers who claim— in very different ways—that a special sort of consciousness is a precondition of being a moral agent are Anscombe, in *Intention,* Sartre, in *The Transcendence of the Ego*, and Harry Frankfurt, in his paper, "Freedom of the Will and the Concept of a Person."[7]

I will argue that the order I have given these six themes is—with one proviso—the order of their dependence. The proviso is that the first three are mutually interdependent; being rational is being Intentional is being the object of a certain stance. These three together are a necessary, but not sufficient condition for exhibiting the form of reciprocity that in turn is a necessary but not sufficient condition for having the capacity for verbal communication, which is the necessary[8] condition for having a special sort of consciousness, which is, as Anscombe and Frankfurt in their different ways claim,[9] a necessary condition of moral personhood.

I have previously exploited the first three themes, rationality, Intentionality and stance, to define not persons, but the much wider class of what I call *Intentional systems,* and since I intend to build on that notion, a brief resume is in order. An Intentional system is a system whose behavior can be (at least sometimes) explained and predicted by relying on ascriptions to the system of *beliefs* and *desires* (and other intentionally characterized features —what I will call *Intentions* here, meaning to include hopes, fears, intentions, perceptions, expectations, etc.) There may be *in every case* other ways of predicting and explaining the behavior of an Intentional system— for instance, mechanistic or physical ways—but the Intentional stance may be the handiest or most effective or in any case a successful stance to adopt, which suffices for the object to be an Intentional system. So defined, Intentional systems are obviously not all persons. We ascribe beliefs and desires to dogs and fish and thereby predict their behavior, and we can even use the procedure to predict the behavior of some machines. For instance, it is a good, indeed the only good, strategy to adopt against a good chess-playing computer. By *assuming* the computer has certain beliefs (or information) and desires (or preference functions) dealing with the chess game in progress, I can calculate—under auspicious circumstances—the computer's most likely next move, *provided I assume the computer deals rationally with these beliefs and desires.* The computer is an Intentional system in these instances not because it has any particular intrinsic features, and not because it really and truly has beliefs and desires (whatever that would be), but just because it succumbs to a certain *stance* adopted toward it, namely the Intentional stance, the stance that proceeds by ascribing Intentional predicates under the usual constraints to the computer, the stance that proceeds by considering the computer as a rational practical reasoner.

It is important to recognize how bland this definition of *Intentional system* is, and how correspondingly large the class of Intentional systems can be. If, for instance, I predict that a particular plant—say a potted ivy—will grow around a corner and up into the light, because it "seeks" the light, and "wants" to get out of the shade it now finds itself in, and "expects" or "hopes" there is light around the corner, I have adopted the Intentional stance toward the plant, and lo and behold, within very narrow limits it works. Since it works, some plants are very low-grade Intentional systems.

The actual utility of adopting the Intentional stance toward plants was brought home to me talking with loggers in the Maine woods. These men invariably call a tree not "it," but "he" and will say of a young spruce "he wants to spread his limbs, but don't let him; then he'll have to stretch up to get his light," or "pines don't like to get their feet wet the way cedars do." You can trick an apple tree into "thinking it's spring" by building a small fire under its branches in the late fall; it will blossom. This way of talking is not just picturesque and is not really superstitious at all; it is simply an efficient way of making sense of, controlling, predicting, and explaining the behavior of these plants in a way that nicely circumvents one's ignorance of the controlling mechanisms. More sophisticated biologists may choose to speak of information transmission from the tree's periphery to other locations in the tree. This is less picturesque, but still Intentional. Complete abstention from Intentional talk about trees can become almost as heroic, cumbersome, and pointless as the parallel strict behaviorist taboo when speaking of rats and pigeons. And even when Intentional glosses on (e.g.) tree-activities are of vanishingly small heuristic value, it seems to me wiser to grant that such a tree is a very degenerate, uninteresting, negligible Intentional system than to attempt to draw a line above which Intentional interpretations are "objectively true."

It is obvious, then, that being an Intentional system is not a sufficient, but is surely a necessary, condition for being a person. Nothing to which we could not successfully adopt the Intentional stance, with its presupposition of rationality, could count as a person. Can we then define persons as a subclass of Intentional systems? At first glance it might seem profitable to suppose that persons are just that subclass of Intentional systems that *really* have beliefs, desires, and so forth, and are not merely *supposed* to have them for the sake of a short cut prediction. But efforts to say what counts as really having a belief (so that no dog, or tree, or computer could qualify) all seem to end by putting conditions on genuine belief that (1) are too strong for our intuitions, and (2) allude to distinct conditions of personhood farther down my list. For instance, one might claim that genuine beliefs are necessarily *verbally expressible* by the believer,[10] or the believer must be *conscious* that he has them, but people seem to have many beliefs that they cannot put into words, and many that they are unaware of having—and in any case I hope to show that the capacity for verbal expression, and capacity for consciousness, find different *loci* in the set of necessary conditions of personhood.

Better progress can be made, I think, if we turn to our fourth theme, reciprocity, to see what kind of definition it could receive in terms of Intentional systems. The theme suggests that a person must be able to reciprocate the stance, which suggests that an Intentional system that itself adopted the Intentional stance toward other objects would meet the test. Let us define a *second-order Intentional system* as one to which we ascribe not only simple beliefs, desires, and other Intentions, but beliefs, desires, and other Intentions *about* beliefs, desires, and other Intentions. An Intentional system S would be a second-order Intentional system if among the ascriptions we make to it are such as *S believes that T desires that p, S hopes that T fears that q,* and reflexive cases like *S believes that S desires that p.* (The importance of the reflexive cases will loom large, not surprisingly, when we turn to those who interpret our sixth conditions as *self*-consciousness. It may seem to some that the reflexive cases make all Intentional systems auto-matically second-order systems and even *n*-order systems, on the grounds that believing that *p* implies believing that you believe that *p* and so forth, but this is a fundamental mistake; the iteration of beliefs and other intentions is never redundant and hence, while some iterations are normal—are to be expected—they are never trivial or automatic.)

Now are human beings the only second-order Intentional systems so far as we know? I take this to be an empirical question. We ascribe beliefs and desires to dogs, cats, lions, birds, and dolphins, for example, and thereby often predict their behavior—when all goes well—but it is hard to think of a case where an animal's behavior was so sophisticated that we would need to ascribe second-order Intentions to predict or explain its behavior. Of course if some version of mechanistic physicalism is true (as I believe), we will never *need* absolutely to ascribe any Intentions to anything, but supposing that for heuristic and pragmatic reasons we were to ascribe Intentions to animals, would we ever feel the pragmatic tug to ascribe second-order Intentions to them? Psychologists have often appealed to a principle known as Lloyd Morgan's Canon of Parsimony, which can be viewed as a special case of Occam's Razor; it is the principle that one should attribute to an organism as little intelligence or consciousness or rationality or mind as will suffice to account for its behavior. This principle can be, and has been, interpreted as demanding nothing short of radical behaviorism,[11] but I think this is a mistake and we can interpret it as the principle requiring us when we adopt the Intentional stance toward a thing to ascribe the simplest, least sophisticated, lowest-order beliefs, desires, and so on, that will account for the behavior. Then we will grant, for instance, that Fido *wants* his supper, and *believes* his master will give him his supper if he begs in front of his master, but we need not ascribe to Fido the further *belief* that his begging induces a *belief* in his master that he, Fido, *wants* his supper. Similarly, my *expectation* when I put a dime in the candy machine does not hinge on a further *belief* that inserting the coin induces the machine to *believe* I *want* some candy. That is, while Fido's begging looks very much like true second-order interacting (with Fido treating his master as an Intentional system), if we suppose that to Fido his master is just a supper machine activated by begging, we will have just as good a predictive ascription, more modest but still, of course, Intentional.

Are dogs, then, or chimps or other "higher" animals, incapable of rising to the level of second-order Intentional systems, and if so why? I used to think the answer was Yes, and I thought the reason was that nonhuman animals lack language, and that language was needed to represent second-order Intentions. In other words, I thought condition four might rest on condition five. I was tempted by the hypothesis that animals cannot, for instance, have second-order beliefs, beliefs about beliefs, for the same reason they cannot have beliefs about Friday, or poetry. Some beliefs can only be acquired, and hence represented, via language.[12] But if it is true that some beliefs cannot be acquired without language, it is false that second-order beliefs are among them, and it is false that nonhumans cannot be second-order Intentional systems. Once I began asking people for examples of nonhuman second-order Intentional systems, I found some very plausible cases. Consider this from Peter Ashley (in a letter):

> One evening I was sitting in a chair at my home, the only chair my dog is allowed to sleep in. The dog was lying in front of me, whimpering. She was getting nowhere in her trying to "convince" me to give up the chair to her. Her next move is the most interesting, nay, the only interesting part of the story. She stood up, and went to the front door where I could still easily see her. She scratched the door, giving me the impression that she had given up trying to get the chair and had decided to go out. However as soon as I reached the door to let her out, she ran back across the room and climbed into her chair, the chair she had "forced" me to leave.

Here it seems we must ascribe to the dog the *intention* that her master believes she *wants* to go out—not just a second-order, but a third-order Intention. The key to the example, what makes it an example of a higher-order Intentional system at work, is that the belief she intends to induce in her master is false. If we want to discover further examples of animals behaving as second-order Intentional systems, it will help to think of cases of deception, where the animal, believing *p*, tries to get another Intentional system to believe not-*p*. Where an animal is trying to induce behavior in another which *true* beliefs about the other's environment would not induce, we cannot "divide through" and get an explanation that cites only first-level Intentions. We can make this point more general before explaining why it is so: where *x* is attempting to induce behavior in *y* which is inappropriate to *y's true* environment and needs but appropriate to *y's perceived* or *believed* envi-ronment and needs, we are forced to ascribe second-order Intentions to *x*. Once in this form, the point emerges as a familiar one, often exploited by critics of behaviorism: one can be a behaviorist in explaining and controlling the behavior of laboratory animals only so long as he can rely on there being no serious dislocation between the actual environment of the experiment and the environment perceived by the animals. A tactic for embarrassing behaviorists in the laboratory is to set up experiments that deceive the subjects: if the deception succeeds, their behavior is predictable from their false *beliefs* about the environment, not from the actual environment. Now a first-order Intentional system is a behaviorist; it ascribes no Intentions to anything. So if we are to have good evidence that some system *S* is *not* a behaviorist—is a second-order Intentional system—it will only be in those cases where behaviorist theories are inadequate to the data, only in those cases where behaviorism would not explain system *S*'s success in manipulating another system's behavior.

This suggests that Ashley's example is not so convincing after all, that it can be defeated by supposing his dog is a behaviorist of sorts. She need not believe that scratching on the door will induce Ashley to believe she wants to go out; she may simply believe, as a good behaviorist, that she has conditioned Ashley to go to the door when she scratches. So she applies the usual stimulus, gets the usual reponse, and that's that. Ashley's case succumbs if this is a standard way his dog has of getting the door opened, as it probably is, for then the more modest hypothesis is that the dog believes her master is conditioned to go to the door when she scratches. Had the dog done something *novel* to deceive her master (like running to the window and looking out, growling suspiciously) then we would have to grant that rising from the chair was no mere conditioned response in Ashley, and could not be "viewed" as such by his dog, but then, such virtuosity in a dog would be highly implausible.

Yet what is the difference between the implausible case and the well-attested cases where a low-nesting bird will feign a broken wing to lure a predator away from the nest? The effect achieved is novel, in the sense that the bird in all likelihood has not repeatedly conditioned the predators in the neighborhood with this stimulus, so we seem constrained to explained the ploy as a bit of genuine deception, where the bird *intends* to induce a false *belief* in the predator. Forced to this interpretation of the behavior, we would be mightily impressed with the bird's ingenuity were it not for the fact that we know such behavior is "merely instinctual." But why does it disparage this trick to call it merely instinctual? To claim it is instinctual is to claim that all birds of the species do it; they do it even when circumstances aren't entirely appropriate; they do it when there are better reasons for staying on the nest; the behavior pattern is rigid, a tropism of sorts, and presumably the controls are generically wired in, not learned or invented.

We must be careful not to carry this disparagement too far; it is not that the bird does this trick "unthinkingly," for while it is no doubt true that she does not in any sense run through an argument or scheme in her head ("Let's see, if I were to flap my wing as if it were broken, the fox would think..."), a man might do something of similar subtlety, and of genuine intelligence, novelty, and appropriateness, and not run through the "conscious thoughts" either. Thinking the thoughts, however that are characterized, is not what makes truly intelligent behavior intelligent. Anscombe says at one point "If [such an expression of reasoning] were supposed to describe actual mental processes, it would, in general, be quite absurd. The interest of the account is that it described an order which is there whenever actions are done with intentions."[13] But the "order is there" in the case of the bird as well as the man. That is, when we ask why birds evolved with this tropism we explain it by noting the utility of having a means of deceiving predators, or inducing false beliefs in them; what must be explained is the provenance of the bird's second-order Intentions. I would be the last to deny or dismiss the vast difference between instinctual or tropistic behavior and the more versatile, intelligent behavior of humans and others, but what I want to insist on here is that if one is prepared to adopt the Intentional stance without qualms as a tool in predicting and explaining behavior, the bird is as much a second-order Intentional system as any man. Since this is so, we should be particularly suspicious of the argument I was tempted to use, viz., that *representations* of second-order Intentions would depend somehow on language.[14] For it is far from clear that all or even any of the beliefs and other Intentions of an Intentional system need be *represented* "within" the system in any way for us to get a purchase on predicting its behavior by *ascribing* such Intentions to it.[15] The situation we elucidate by citing the bird's desire to induce a false belief in the predator seems to have no room or need for a representation of this sophisticated Intention in any entity's "thoughts" or "mind," for neither the bird, nor evolutionary history, nor Mother Nature need think these thoughts for our explanation to be warranted.

Reciprocity, then, provided we understand it by merely the capacity in Intentional systems to exhibit higher-order Intentions, while it depends on the first three conditions, is independent of the fifth and sixth. Whether this notion does justice to the reciprocity discussed by other writers will begin to come clear only when we see how it meshes with the last two conditions. For the fifth condition, the capacity for verbal communication, we turn to Grice's theory of meaning. Grice attempts to define what he calls nonnatural meaning, an utterer's meaning something by uttering something, in terms of the *intentions* of the utterer.

His initial definition is as follows[16]:

"U meant something by uttering x" is true if, for some audience A, U uttered x intending
(1) A to produce a particular response *r*.
(2) A to think (recognize) that U intends (1).
(3) A to fulfill (1) on the basis of his fulfillment of (2).

Notice that intention (2) ascribes to U not only a second-, but a third-order Intention: U must *intend* that A *recognize* that U *intends* that A produce *r*. It matters not at all that Grice has been forced by a series of counterexamples to move from this initial definition to much more complicated versions, for they all reproduce the third-order Intention of (2). Two points of great importance to us emerge from Grice's analysis of nonnatural meaning. First, since nonnatural meaning, meaning something by saying something, must be a feature of any true verbal communication, and since it depends on third-order Intentions on the part of the utterer, we have our case that condition five rest on condition four and not vice versa. Second, Grice shows us that mere *second*-order Intentions are not enough to provide genuine reciprocity; for that, *third*-order Intentions are needed. Grice introduces condition (2) in order to exclude such cases as this: I leave the china my daughter has broken lying around for my wife to see. This is not a case of meaning something by doing what I do intending what I intend, for though I am attempting thereby to induce my wife to believe something about our daughter (a second-order intention on my part), success does not depend on her recognizing this intention of mine, or recognizing my intervention or existence at all. There has been no real encounter, to use Erving Goffman's apt term, between us, no mutual recognition.

There must be an encounter between utterer and audience for utterer to mean anything, but encounters can occur in the absence of non-natural meaning (witness Ashley's dog), and ploys that depend on third-order Intentions need not involve encounters (e.g., A can intend that B believe that C desires that p). So third-order Intentions are a necessary, but not sufficient condition for encounters which are a necessary, but not sufficient condition for instances of nonnatural meaning, that is, instances of verbal communication.

It is no accident that Grice's cases of nonnatural meaning fall into a class whose other members are cases of deception or manipulation. Consider, for instance, Searle's ingenious counterexample to one of Grice's formulations: the American caught behind enemy lines in World War II Italy who attempts to deceive his Italian captors into concluding he is a German officer by saying the one sentence of German he knows: *"Kennst du das Land, wo die Zitronen blühen?* [17] As Grice points out, these cases share with cases of nonnatural meaning a reliance on or exploitation of the rationality of the victim. In these cases success hinges on inducing the victim to embark on a chain of reasoning to which one contributes premises directly or indirectly. In deception, the premises are disbelieved by the supplier; in normal communication, they are believed. Communication, in Gricean guise, appears to be a sort of collaborative manipulation of audience by utterer; it depends, not only on the rationality of the audience who must sort out the utterer's intentions, but on the audience's *trust* in the utterer. Communication, as a sort of manipulation, would not work, given the requisite rationality of the audience, unless the audience's trust in the utterer were *well-grounded* or reasonable. Thus the *norm* for utterance is sincerity; were utterances not normally trustworthy, they would fail of their purpose.[18]

Lying, as a form of deception, can only work against a background of truth-telling, but other forms of deception do not de–pend on the trust of the victim. In these cases, success depends on the victim being *quite* smart, but not quite smart enough. Stupid poker players are the bane of clever poker players, for they fail to see the bluffs and ruses being offered them. Such sophisticated deceptions need not depend on direct encounters. There is a book on how to detect fake antiques (which is also, inevitably, a book on how to *make* fake antiques) which offers this sly advice to those who want to fool the "expert" buyer; once you have completed your table or whatever (having utilized all the usual means of simulating age and wear,) take a modern electric drill and drill a hole right through the piece in some conspicuous but perplexing place. The would-be buyer will argue: no one would drill such a disfiguring hole without a reason (it can't be supposed to look "authentic" in any way) so it must have served a purpose, which means this table must have been in use in someone's home; since it was in use in someone's home, it was not made expressly for sale in this antique shop...therefore it is authentic. Even if this "conclusion" left room for lingering doubts, the buyer will be so preoccupied dreaming up uses for that hole it will be months before the doubts can surface.

What is important about these cases of deception is the fact that just as in the case of the feigning bird, success does not depend on the victim's *consciously entertaining* these chains of reasoning. It does not matter if the buyer just notices the hole and "gets a hunch" the piece is genuine. He *might* later accept the reasoning offered as his "rationale" for finding the piece genuine, but he might deny it, and in denying it, he might be deceiving himself, even though the thoughts never went through his head. The chain of reasoning explains why the hole works as it does (if it does), but as Anscombe says, it need not "describe actual mental processes," if we suppose actual mental processes are conscious processes or events. The same, of course, is true of Gricean communications; neither the utterer nor the audience need consciously entertain the complicated Intentions he outlines, and what is a bit surprising is that no one has ever used this fact as an objection to Grice. Grice's conditions for meaning have been often criticized for falling short of being suffic-ient, but there seems to be an argument not yet used to show they are not even necessary. Certainly few people ever consciously framed those ingenious intentions before Grice pointed them out, and yet people had been communicating for years. Before Grice, were one asked: "Did you intend your audience to recognize your intention to provoke that response in him?" one would most likely have retorted: "I intended nothing so devious. I simply intended to inform him that I wouldn't be so devious. I simply intended to inform him that I wouldn't be home for supper" (or whatever). So it seems that if these complicated intentions underlay our communicating all along, they must have been unconscious intentions. Indeed, a perfectly natural way of responding to Grice's papers is to remark that *one was not aware* of doing these things when one communicated. Now Anscombe has held, very powerfully, that such a response establishes that the action under that description was not intentional.[19] Since one is not *aware* of these intentions in speaking, one cannot be speaking *with* these intentions.

Why has no one used this argument against Grice's theory? Because, I submit, it is just too plain that Grice is on to something, that Grice is giving us necessary conditions for nonnatural meaning. His analysis illuminates so many questions. Do we communicate with computers in Fortran? Fortran seems to be a language; it has a grammar, a vocabulary, a semantics. The transactions in Fortran between man and machine are often viewed as cases of *man communicating with machine,* but such transactions are pale copies of human verbal communication precisely because the Gricean conditions for nonnatural meaning have been bypassed. There is no room for them to apply. Achieving one's ends in transmitting a bit of Fortran to the machine does not hinge on getting the machine to recognize one's intentions. This does not mean that all communications with computers in the future will have this shortcoming (or strength, depending on your purposes), but just that we do not communicate now, in the strong (Gricean) sense, with computers.[20]

If we are not about to abandon the Gricean model, yet are aware of no such intentions in our normal conversation, we shall just have to drive these intentions underground, and call them unconscious or preconscious intentions. They are intentions that exhibit "an order which is there" when people communicate, intentions of which we are not normally aware, and intentions that are a precondition of verbal communication.[21]

We have come this far without having to invoke any sort of consciousness at all, so if there is a dependence between consciousness or self-consciousness and our other conditions, it will have to be consciousness depending on the others. But to show this, I must first show how the first five conditions by themselves might play a role in ethics, as suggested by Rawls's theory of justice. Central to Rawls's theory is his setting up of an idealized situation, the "original position," inhabited by idealized persons and deriving from this idealization the first principles of justice that generate and illuminate the rest of his theory. What I am concerned with now is neither the content of these principles nor the validity of their derivation, but the nature of Rawls's tactic. Rawls supposes that a group of idealized persons, defined by him as rational, self-interested entities, make calculations under certain constraints about the likely and possible interactive effects of their individual and antagonistic interests (which will require them to frame higher-order Intentions, for example, beliefs about the desires of others, beliefs about the beliefs of others about their own desires, and so forth). Rawls claims these calculations have an optimal "solution" that it would be reasonable for each self-interested person to adopt as an alternative to a Hobbesian state of nature. The solution is to agree with his fellows to abide by the principles of justice Rawls adumbrates. What sort of a proof of the principles of justice would this be? Adopting these principles of justice can be viewed, Rawls claims, as the solution to the "highest order game" or "bargaining problem." It is analogous to derivations of game theory, and to proofs in Hintikka's epistemic logic,[22] and to a "demonstration" that the chess-playing computer will make a certain move because it is the most rational move given its information about the game. All depend on the assumption of ideally rational calculators and hence their outcomes are intrinsically normative. Thus I see the derivations from Rawls's original position as continuous with the deductions and extrapolations encountered in more simple uses of the Intentional stance to understand and control the behavior of simpler entities. Just as truth and consistency are norms for belief,[23] and sincerity is the norm for utterance, so, if Rawls is right, justice as he defines it is the norm for interpersonal interactions. But then, just as part of our warrant for considering an entity to have any beliefs or other Intentions is our ability to construe the entity as *rational,* so our grounds for considering an entity a person include our ability to view him as abiding by the principles of justice. A way of capturing the peculiar status of the concept of a person as I think it is exploited here would be to say that while Rawls does not intend at all to argue that justice is the inevitable result of *human* interaction, he does argue in effect that it is the inevitable result of *personal* interaction. That is, the concept of a person is itself inescapably normative or idealized; to the extent that justice does not reveal itself in the dealings and interactions of creatures, to that extent they are not persons. And once again we can see that there is "an order which is there" in a just society that is independent of any actual episodes of conscious thought. The existence of just practices and the "acknowledgment" implicit in them does not depend on anyone ever consciously or deliberately going through the calculations of the idealized original position, consciously arriving at the reciprocal agreements, consciously adopting a stance toward others.

> To recognize another as a person one must respond to him and act towards him in certain ways; and these ways are intimately connected with the various prima facie duties. Acknowledging these duties in some drgree, and so having the elements of morality, is not a matter of choice or of intuiting moral qualities or a matter of the expression of feelings or attitudes...it is simply the pursuance of one of the forms of conduct in which the recognition of others as persons is manifested.[24]

The importance of Rawls's attempt to derive principles of justice from the "original position" is, of course, that the whole outcome is recognizable as a *moral* norm, it is not *derived* as a moral norm. Morality is not presupposed of the parties in the original position. But this means that the derivation of the norm does not of itself give us any answer to the questions of when and why we have the right to hold persons *morally* responsible for deviations from that norm. Here Anscombe provides help and at the same time introduces our sixth condition. *If I am to be held responsible for an action* (a bit of behavior of mine under a particular description), I must have been *aware* of that action under that description.[25] Why? Because only if I was aware of the action can I say what I was about and participate from a privileged position in the question-and-answer game of giving reasons for my actions. (If I am not in a privileged position to answer questions about the reasons for my actions, there is no special reason to ask me.) And what is so important about being able to participate in this game is that only those capable of participating in reason-giving can be argued into, or argued out of, courses of action or attitudes, and if one is incapable of "listening to reason" in some matter, one cannot be held responsible for it. The capacities for verbal communication and for awareness of one's actions are thus essential in one who is going to be amenable to argument or persuasion, and such persuasion, such reciprocal adjustment of interests achieved by mutual exploitation of rationality, is a feature of the optimal mode of personal interaction.

This capacity for participation in mutual persuasion provides the foundation for yet another condition of personhood recently exposed by Harry Frankfurt.[26] Frankfurt claims that persons are the subclass of Intentional systems capable of what he calls *second-order volitions*. Now at first this looks just like the class of second-order Intentional systems, but it is not, as we shall see.

> Besides wanting, choosing, and being moved to do this or that, men may also want to have (or not to have) certain desires and motives. They are capable of wanting to be different, in their preferences and purposes, from what they are....No animal other than man, however, appears to have the capacity for reflective self-evaluation that is manifested in the formation of second-order desires. (p. 7)

Frankfurt points out that there are cases in which a person might be said to want to have a particular desire even though he would not want that desire to be effective for him, to be "his will." (One might, for instance, want to desire heroin just to know what it felt like to desire heroin, without at all wanting this desire to become one's effective desire.) In more serious cases one wants to have a desire one currently does not have and wants this desire to become one's will. These cases Frankfurt calls second-order volitions, and it is having these, he claims, that is "essential to being a person" (p. 10). His argument for this claim, which I will try not to do justice to here, proceeds from an analysis of the distinction between having freedom of action and having freedom of the will. One has freedom of the will, on his analysis, only when one can have the will one wants, when one's second-order volitions can be satisfied. Persons do not always have free will and under some circumstances can be responsible for actions done in the absence of freedom of the will, but a person always must be an "entity for whom the freedom of its will may be a problem" (p. 14)—that is, one capable of framing second-order volitions, satisfiable or not. Frankfurt introduces the marvelous term "wanton" for those "who have first-order desires but...no second-order volitions." (Second-order volitions for Frankfurt are all, of course, *reflexive* second-order desires.) He claims that our intuitions support the opinion that all nonhuman animals, as well as small children and some mentally defective people, are wantons and I for one can think of no plausible counterexamples. Indeed, it seems a strength of his theory, as he claims, that human beings—the only persons we recognize—are distinguished from animals in this regard. But what should be so special about second-order volitions? Why are they, among higher-order Intentions, the peculiar province of persons? Because, I believe, the "reflective self-evaluation" Frankfurt speaks of is, and must be, genuine self-consciousness, which is achieved only by adopting toward *oneself* the stance not simply of communicator but of Anscombian reason-asker and persuader. As Frankfurt points out, second-order desires are an empty notion unless one can *act* on them, and acting on a second-order desire must be logically distinct from acting on its first-order component. Acting on a second-order desire, doing something to bring it about that one acquires a first-order desire, is acting upon oneself just as one would act upon another person: one schools oneself, one offers oneself persuasions, arguments, threats, bribes, in the hopes of inducing oneself to acquire the first-order desire.[27] One's stance toward oneself *and access to oneself* in these cases is essentially the same as one's stance toward and access to another. One must *ask oneself* what one's desires, motives, reasons really are, and only if one can say, can become aware of one's desires, can one be in a position to induce oneself to change.[28] Only here, I think, is it the case that the "order which is there" cannot be there unless it is there in episodes of conscious thought, in a dialog with oneself.[29]

Now, finally, why are we not in a position to claim that these necessary conditions of moral personhood are also sufficient? Simply because the concept of a person is, I have tried to show, inescapably normative. Human beings or other entities only can aspire to being approximations of the ideal, and there can be no way to set a "passing grade" that is not arbitrary. Were the six conditions (strictly interpreted) considered sufficient they would not ensure that any actual entity was a person, for nothing ever would fulfill them. The moral notion of a person and the metaphysical notion of person are not separate and distinct concepts, but just two different and unstable resting points on the same continuum. This relativity infects the satisfaction of conditions of personhood at every level. There is no objectively satisfiable sufficient condition for an entity's *really* having beliefs, and as we uncover apparent irrationality under an Intentional interpretation of an entity, our grounds for ascribing any beliefs at all wanes, especially when we have (what we always *can* have in principle) a non-Intentional, mechanistic account of the entity. In just the same way our assumption that an entity is a person is shaken precisely in those cases where it matters: when wrong has been done and the question of responsibility arises. For in these cases the grounds for saying that the person is culpable (the evidence that he did wrong, was aware he was doing wrong, and did wrong of his own free will) are in themselves grounds for doubting that it is a person we are dealing with at all. And if it is asked what could *settle* our doubts, the answer is: nothing. When such problems arise we cannot even tell in our own cases if we are persons.

Notes

[1]*See* my "Mechanism and Responsibility," in T. Honderich, ed., *Essays on Freedom of Action* (London: Routledge & Kegan Paul, 1973).

[2]In "Justice as Reciprocity," a revision of "Justice as Fairness" printed in S. Gorovitz, ed., *Utilitarianism* (Indianapolis: Bobbs Merrill, 1971), Rawls allows that the persons in the original position may include "nations, provinces, business firms, churches, teams, and so on. The principles of justice apply to conflicting claims made by persons of all these separate kinds. There is, perhaps, a certain logical priority to the case of human individuals" (p. 245). In *A Theory of Justice* (Cambridge, MA., Harvard University Press, 1971), he acknowledges that parties in the original position may include associations and other entities not human individuals (e.g., p. 146), and the apparent interchangeability of "parties in the original position" and "persons in the original position" suggests that Rawls is claiming that for some moral concept of a person, the moral person is *composed* of metaphysical persons who may or may not themselves be moral persons.

[3]Setting aside Rawls's possible compound moral persons. For more on compound persons, see Amelie Rorty, "Persons, Policies, and Bodies," *International Philosophical Quarterly* **XIII**, 1 (March 1973).

[4]J. Hintikka, *Knowledge and Belief* (Ithaca: Cornell University Press, 1962).

[5]P. F. Strawson, *Individuals* (London: Methuen, 1959), pp. 101–102. It has often been pointed out that Strawson's definition is obviously much too broad, capturing all sentient, active creatures. See, e.g., H. Frankfurt, "Freedom of the

Will and the Concept of a Person," *Journal of Philosophy* (January 14, 1971). It can also be argued (and I would argue) that states of consciousness are only a proper subset of psychological or Intentionally characterized states, but I think it is clear that Strawson here means to cast his net wide enough to include psychological states generally.

[6]D. M. MacKay, "The use of behavioral language to refer to mechanical processes," *British Journal of Philosophy of Science* (1962), 89–103; P. F. Strawson, "Freedom and Resentment," *Proceedings of the British Academy* (1962), reprinted in Strawson, ed., *Studies in the Philosophy of Thought and Action* (Oxford, 1968); A. Rorty, "Slaves and Machines," *Analysis* (1962); H. Putnam, "Robots: Machines or Artificially Created Life?" *Journal of Philosophy* (November 12, 1964); W. Sellars, "Fatalism and Determinism," K. Lehrer, ed., *Freedom and Determinism* (New York: Random House, 1966); A. Flew, "A Rational Animal," J. R. Smythies, ed., *Brain and Mind* (London: Routledge & Kegan Paul, 1968); T. Nagel, "War and Massacre," *Philosophy and Public Affairs* (Winter 1972); D. Van de Vate, "The Problem of Robot Consciousness," *Philosophy and Phenomenological Research* (December 1971); my "Intentional Systems," *Journal of Philosophy* (February 25, 1971).

[7]H. Frankfurt, "Freedom of the Will and the Concept of a Person," *op. cit.*

[8]And sufficient, but I will not argue it here. I argue for this in *Content and Consciousness* (London: Routledge & Kegan Paul, 1969), and more recently and explicitly in my "Reply to Arbib and Gunderson," APA Eastern Division Meetings, December 29, 1972.

[9]I will not discuss Sartre's claim here.

[10]Cf. Bernard Williams, "Deciding to Believe," in H. E. Kiefer and M. K. Munitz, eds., *Language, Belief and Metaphysics* (New York: New York University Press, 1970).

[11]E.g., B. F. Skinner, " Behaviorism at Fifty," in T. W. Wann, ed., *Behaviorism and Phenomenology* (Chicago: University of Chicago Press, 1964).

[12]For illuminating suggestions on the relation of language to belief and rationality, *see* Ronald de Sousa, "How to Give a Piece of Your Mind; or, a Logic of Belief and Assent," *Review of Metaphysics* (September 1971).

[13]G. E. M. Anscombe, *Intention* (Oxford: Blackwell, 1957), p. 80.

[14]Cf. Ronald de Sousa, "Self-Deception," *Inquiry* 13, (1970), esp. p. 317.

[15]I argue this in more detail in "Brain Writing and Mind Reading," in K. Gunderson, ed., *Language, Mind, and Knowledge* (Minneapolis: University of Minnesota Press, 1975), and in my "Reply to Arbib and Gunderson."

[16]The key papers are "Meaning," *Philosophical Review* (July 1957), and "Utterer's Meaning and Intentions," *Philosophical Review* (April 1969). His initial formulation, developed in the first paper, is subjected to a series of revisions in the second paper, from which this formulation is drawn (p. 151).

[17]John Searle, "What is a Speech Act ?" in *Philosophy in America*, Max Black, ed. (London: Allen & Unwin, 1965), discussed by Grice in "Utterer's Meaning and Intentions," p. 160.

[18]Cf. "Intentional Systems," pp. 102–103.

[19]G. E. M. Anscombe, *Intention,* p. 11.

[20]It has been pointed out to me by Howard Friedman that many current Fortran compilers which "correct " operator input by inserting "plus" signs and parentheses, etc., to produce well-formed expressions arguably meet Grice's criteria, since within a very limited sphere, they diagnose the "utterer's" intentions and proceed on the basis of this diagnosis. But first it should be noted that the machines to date can diagnose only what might be called the operator's syntactical intentions, and second, these machines do not seem to meet Grice's subsequent and more elaborate definitions, not that I wish to claim that no computer could.

[21]In fact, Grice is describing only a small portion of the order which is there as a precondition of normal personal interaction. An analysis of higher order Intentions on a broader front is to be found in the works of Erving Goffman, especially in *The Presentation of Self in Everyday Life* (Garden City: Doubleday, 1959).

[22]*See* Hintikka, *Knowledge and Belief,* p. 38.

[23]*See* Dennett, "Intentional Systems," pp. 102–103.

[24]J. Rawls, "Justice as Reciprocity, " p. 259.

[25]I can be held resposible for events and states of affairs that I was not aware of and ought to have been aware of, but these are not intentional actions. In these cases I am responsible for these further matters in virtue of being responsible for the foreseeable consequences of actions—including acts of omission—that I was aware of.

[26]H. Frankfurt, "Freedom of the Will and the Concept of a Person." Frankfurt does not say whether he conceives his condition to be merely a necessary or also a sufficient condition of moral personhood.

[27]It has been brought to my attention that dogs at stud will often engage in masturbation, in order, apparently, to *increase their desire* to copulate. What makes these cases negligible is that even supposing the dog can be said to act on a desire to strengthen a desire, the effect is achieved in a non-intentional ("purely physiological") way; the dog does not appeal to or exploit his own rationality in achieving his end. (As if the only way a person could act on a second-order volition were by taking a pill or standing on his head, etc.).

[28]Margaret Gilbert, in "Vices and self-knowledge," *Journal of Philosophy* (August 5, 1971), p. 452, examines the implications of the fact that "when, and only when, one believes that one has a given trait can one decide to change out of it."

[29]Marx, in *The German Ideology,* says: "Language, like consciousness, only arises from the need, the necessity, of intercourse with other men....Language is as old as consciousness, language is practical consciousness." And Nietzsche, in *The Joyful Wisdom,* says: "For we could in fact think, feel, will, and recollect, we could likewise 'act ' in every sense of the term, and nevertheless nothing of it at all need necessarily 'come into consciousness' (as one says metaphorically;....*What* then is *the purpose* of consciousness generally, when it is in the main *superflous?*—Now it seems to me, if you will hear my answer and its perhaps extravagant supposition, that the subtlety and strength of consciousness are always in proportion to the *capacity for communication* of a man (or an animal), the capacity for communi-

cation in its turn being in proportion to the *necessity for communication*....In short, the development of speech and the development of consciousness (not of reason, but of reason becoming self-conscious) go hand in hand."

Medicine and the Concept of Person

H. Tristram Engelhardt, Jr.

Recent advances in medicine and the biomedical sciences have raised a number of ethical issues that medical ethics or, more broadly, bioethics have treated. Ingredient in such considerations, however, are fundamentally conceptual and ontological issues. To talk of the sanctity of life, for example, presupposes that one knows (1) what life is, and (2) what makes for its sanctity. More importantly, to talk of the the rights of persons presupposes that one knows what counts as a person. In this paper I will provide an examination of the concept of person and will argue that the terms "human life" and even "human person" are complex and heterogeneous terms. I will hold that human life has more than one meaning and I will then indicate how the recognition of these multiple meanings has important implications for medicine.

Kinds of Life and Sanctity of Life

Whatever is meant by life's being sacred, it is rarely held that all life is equally sacred. Most people would find the life of bacteria, for example, to be less valuable or sacred than the life of fellow humans. In fact, there appears to be a spectrum of increasing value to life (I will presume that the term sanctity of life signifies that life has either special values or rights). All else being equal, plants seem to be valued less than lower animals, lower animals less than higher animals (such as primates other than humans), and humans are usually held to have the highest value. Moreover, distinctions are made with respect to humans. Not all human life has the same sanctity. The

169

issue of brain death, for example, turns on such a distinction. Brain dead, but otherwise alive, human beings do not have the sanctity of normal adult human beings. That is, the indices of brain death have been selected in order to measure the death of a person. As a legal issue, it is a question of when a human being ceases to be a person before the law. In a sense, the older definition of death measured the point at which organismic death occurred, when there was a complete cessation of vital functions.[1] The life of the human organism was taken as a necessary condition from being a person, and, therefore, such a definition allowed one to identify cases in which humans ceased to be persons.

The brain-oriented concept of death is more directly concerned with human *personal* life.[2] It makes three presuppositions: (1) that being a person involves more than mere vegetative life, (2) that merely vegetative life may have value but it has no rights, (3) that a sensory-motor organ such as the brain is a necessary condition for the possibility of experience and action in the world, that is, for being a person living in the world. Thus in the absence of the possibility of brain-function, one has the absence of the possibility of personal life—that is, the person is dead. Of course, the presence of some brain activity (or more than vegetative function) does not imply the presence of a person—a necessary condition for the life of a person. The brain-oriented concept of death is of philosophical significance, for, among other things, it implies a distinction between human biological life and human personal life, between the life of a human organism and the life of a human person. That human biological life continues after brain death is fairly clear: the body continues to circulate blood, the kidneys function; in fact, there is no reason why the organism would not continue to be cross-fertile (e.g., produce viable sperm) and, thus, satisfy yet one more criterion for biological life. Such a body can be a biologically integrated reproductive unit even if the level of integration is very low. And, if such a body is an instance of human biological but not human personal life, then it is open to use merely as a subject of experimentation without the constraints of a second status as a person. Thus Dr. Willard Gaylin has argued that living but brain-dead bodies could provide an excellent source of subjects for medical experimentation and education[3] and recommends "sustaining life in the brain dead."[4] To avoid what would otherwise be an oxymoronic position, he is legitimately pressed to distinguish, as he does in fact, between "aliveness" and "personhood,"[5] or, to use more precise terminology, between human biological and human personal life. In short, a distinction between the status of human biological and personal life is presupposed.

We are brought then to a set of distinctions: first, human life must be distinguished as human personal and human biological life. Not all instances of human biological life are instances of human personal life. Brain dead (but otherwise alive) human beings, human gametes, cells in human cell cultures, all count as instances of human biological life. Further, not only are some humans not persons, there is no reason to hold that all persons are humans, as the possibility of extraterrestrial self-conscious life suggests.

Second, the concept of the sanctity of life comes to refer in different ways to the value of biological life and the dignity of persons. Probably much that is associated with arguments concerning the sanctity of life really refers to the dignity of the life of persons. In any event, there is no unambiguous sense of being simply "pro-life" or a defender of the sanctity of life—one must decide what sort of life one wishes to defend and on what grounds. To begin with, the morally significant difference between biological and personal life lies in the fact, to use Kant's idiom, that persons are ends in themselves. Rational, self-conscious agents can make claims to treatment as ends in themselves because they can experience themselves, can know that they experience themselves, and can determine and control the circumstances of such experience. Self-conscious agents are self-determining and can claim respect as such. That is, they can claim the right to be respected as free agents. Such a claim is to the effect that self-respect and mutual respect turn on self-determination, on the fact that self-conscious beings are necessary for the existence of a moral order—a kingdom of ends, a community based on mutual self-respect, not force. Only self-conscious agents can be held accountable for their actions and thus be bound together solely in terms of mutual respect of each other's autonomy.

What I intend here is no more than an exegesis of what we could mean by "respecting persons." Kant, for example, argued that rational beings are "persons, because their very nature [as rational beings] points them out as ends in themselves."[6] In this fashion, Kant developed a distinction between things that have only "a worth *for us"* and persons "whose existence is an end in itself."[7] As a result, Kant drew a stark and clear distinction between persons and nonpersons. "A person is [a] subject whose actions are capable of being imputed [that is, one who can act responsibly]. Accordingly, moral personality is nothing but the freedom of a rational being under moral laws (whereas psychological personality is merely the capacity to be conscious of the identity of one's self in the various conditions of one's existence)...[In contrast], a thing is that which is not capable of any imputation [that is, of acting responsibly]."[8] To be respected as a moral agent is precisely to be respected as a free self-conscious being capable of being blamed and praised, of being held responsible for its actions. The language of respect in the sense of recognizing others as free to determine themselves (i.e., as ends in themselves) rather than as beings to be determined by others (i.e., to be used as means; instruments to goods and values) turns upon acknowledging others as free, as moral agents.

This somewhat obvious exegesis (or tautological point) is an account of the nature of the language of obligation. Talk of obligation functions (1) to remind us that certain actions cannot be reconciled with the notion of a moral community, and (2) to enjoin others to pursue particular values or goods. The only actions that strictly contradict the notion of a moral community are those that are incompatible with the notion of such a community—actions that treat moral agents as if they were objects. Morality as mutual respect (i.e., more than conjoint pursuit of particular goods or goals) can be consistently pursued only if persons in the strict sense (i.e., self-conscious agents, entities able to be self-legislative) are treated with respect. However, no particular ranking of values need be presupposed. Though we may treat other entities with a form or respect, that respect is never central to the notion of a community of moral agents. Insofar as we identify persons with moral agents, we exclude from the range of the concept person those entities which are not self-conscious. Which is to say, only those beings are unqualified bearers of rights and duties who can both claim to be acknowledged as having a dignity beyond a value (i.e., as being ends in themselves), and can be responsible for their actions. Of course, this strict sense of person is not unlike that often used in the law.[9] And, as Kant suggests in the passage above, it requires as well an experience of self-identity through time.

It is only respect for persons in this strict sense that cannot be violated without contradicting the idea of a moral order in the sense of living with others on the basis of mutual respect. The point to be emphasized is a distinction between value and dignity, between biological life and personal life. These distinctions provide a basis for the differentiation between biological or merely animal life, and personal life, and turn on the rather commonsense criterion of respect being given that which can be respected—that is, blamed or praised. Moral treatment comes to depend, not implausibly, on moral agency. The importance of such distinctions for medicine is that they can be employed in treating medical ethical issues. As arguments, they are attempts to sort out everyday distinctions between moral agents, other animals, and just plain things. They provide a conceptual apparatus based on the meaning of obligations as respect due that which can have obligations.

The distinctions between human biological life and human personal life, and between the value of human biological life and the dignity of human personal life, involve a basic conceptual distinction that modern medical science presses as an issue of practical importance. Medicine, after all, is not merely the enterprise of preserving human life—if that were the case, medicine would confuse human cell cultures with patients who are persons. In fact, a maxim "to treat patients as persons" presupposes that we do or can indeed know who the persons are. These distinctions focus not only on the definition of death, but on the question of abortion as well: issues that turn on when persons end and when they begin. In the case of the definition of death, one is saying that even though genetic continuity, organic function, and reproductive capability may extend beyond brain death, personal life does not. Sentience in an appropriate embodiment is a necessary condition for being a person.[10] One, thus, finds that persons die when this embodiment is undermined.

With regard to abortion, many have argued similarly that the fetus is not a person, though it is surely an instance of human biological life. Even if the fetus is a human organism that will probably be genetically and organically continuous with a human person, it is not yet such a person.[11] Simply put, fetuses are not rational, self-conscious beings—that is, given a strict definition of persons, fetuses do not qualify as persons. One sees this when comparing talk about dead

men with talk about fetuses. When speaking of a dead man, one
knows of whom one speaks, the one who died, the person whom one
knew before his death. But in speaking of the fetus, one has no such
person to whom one can refer. There is not yet a person, a "who," to
whom one can refer in the case of the fetus (compare: one can keep
promises to dead men but not to men yet unborn). In short, the fetus
in no way singles itself out as, or shows itself to be, a person. This
conclusion has theoretical advantages, since many zygotes never
implant and some divide into two.[12] It offers as well a moral
clarification of the practice of using intrauterine contraceptive
devices and abortion. Whatever these practices involve, they do not
involve the taking of the life of a person.[13] This position in short
involves recurring to a distinction forged by both Aristotle and St.
Thomas—between biological life and personal life,[14] between life
that has value and life that has dignity.

But this distinction does too much, as the arguments by Michael
Tooley on behalf of infanticide show.[15] By the terms of the argument, infants, as well as fetuses, are not persons—thus, one finds
infants as much open to infanticide as fetuses are left open to abortion. The question then is whether one can recoup something for
infants or perhaps even for fetuses. One might think that a coun-
terargument, or at least a mitigating argument, could be made on the
basis of potentiality—the potentiality of infants or the potentiality of
fetuses. That argument, though, fails because one must distinguish
the potentialities of a person from the potentiality to become a
person. If, for example, one holds that a fetus has the potentiality of
a person, one begs the very question at issue—whether fetuses are
persons. But, on the other hand, if one succeeds in arguing that a
fetus or infant has the potentiality to become a person, one has conceded the point that the fetus or infant is not a person. One may
value a dozen eggs or a handful of acorns because they can become
chickens or oak trees. But a dozen eggs is not a flock of chickens, a
handful of acorns is not a stand of oaks. In short, the potentiality of
X's to become Y's may cause us to value X's very highly because Y's
are valued very highly, but until X's are Y's they do not have the
value of Y's.[16]

Which is to say, given our judgments concerning brain dead humans and concerning zygotes, embryos, and fetuses, we are left in a
quandary with regard to infants. How, if at all, are we to understand

them to be persons, beings to whom we might have obligations? One should remember that these questions arise against the backdrop of issues concerning the disposition of deformed neonates—whether they should all be given maximal treatment, or whether some should be allowed to die, or even have their deaths expedited.[17]

In short, though we have sorted out a distinction between the value of human biological life and the dignity of human personal life, this distinction does not do all we want, or rather it may do too much. That is, it goes against an intuitive appreciation of children, even neonates, as not being open to destruction on request. We may not in the end be able to support that intuition, for it may simply be a cultural prejudice; but I will now try to give a reasonable exegesis of its significance.

Two Concepts of Person

I shall argue in this section that a confusion arises out of a false presupposition that we have only one concept of person: we have at least two concepts (probably many more) of person. I will restrict myself to examining the two that are most relevant here. First, there is the sense of person that we use in identifying moral agents: individual, living bearers of rights and duties. That sense singles out entities who can participate in the language of morals, who can make claims and have those claims respected: the strict sense we have examined above. We would, for example, understand "person" in this sense to be used properly if we found another group of self-conscious agents in the universe and called them persons even if they were not human, though it is a term that usually applies to normal adult humans. This sense of person I shall term the strict sense, one which is used in reference to self-conscious, rational agents. But what of the respect accorded to infants and other examples of non-self-conscious or not-yet-self-conscious human life? How are such entities to be understood?

A plausible analysis can, I believe, be given in terms of a second concept or use of person—a social concept or social role of person that is invoked when certain instances of human biological life are treated as if they were persons strictly, even though they are not. A good example is the mother–child or parent–child relationship in which the infant is treated as a person even though it is not one strictly. That is, the infant is treated as if it had the wants and desires of a person—its cries are treated as a call for food, attention, care, and so on, and the infant is socialized, placed within a social structure, the family, and becomes a child. The shift is from merely biological to social significance. The shift is made on the basis that the infant is a human and is able to engage in a minimum of social interaction. With regard to the latter point, severely anencephalic infants may not qualify for the role *person* just as brain dead adults would fail to qualify; both lack the ability to engage in minimal social interaction.[18] This use of person is, after all, one employed with instances of human biological life that are enmeshed in social roles as if they were persons. Further, one finds a difference between the biological mother-fetus relation and the social mother-child relation. The first relation can continue whether or not there is social recognition of the fetus, the second cannot. The mother-child relation is essentially a social practice.[19]

This practice can be justified as a means of preserving trust in families, of nurturing important virtues of care and solicitude towards the weak, and of assuring the healthy development of children. Further, it has a special value because it is difficult to determine specifically when in human ontogeny persons strictly emerge. Socializing infants into the role *person* draws the line conservatively. Humans do not become persons strictly until some time after birth. Moreover, there is a considerable value in protecting anything that looks and acts in a reasonably human fashion, especially when it falls within an established human social role as infants do within the role *child*. This ascription of the role *person* constitutes a social practice that allows the rights of a person to be imputed to forms of human life that can engage in at least a minimum of social interaction. The interest is in guarding anything that could reasonably play the role *person* and thus to strengthen the social position of persons generally.

The social sense of person appears as well to structure the treatment of the senile, the mentally retarded, and the otherwise severely mentally infirm. Though they are not moral agents, persons strictly, they are treated as if they were persons. The social sense of person identifies their place in a social relationship with persons strictly. It is, in short, a practice that gives to instances of human biological life the status of persons. Unlike persons strictly, who are bearers of both rights and duties, persons in the social sense have rights but no duties. That is, they are not morally responsible agents, but are treated with respect (i.e., rights are imputed to them) in order to establish a practice of considerable utility to moral agents: a society where kind treatment of the infirm and weak is an established practice. The central element of the utility of this practice lies in the fact that it is often difficult to tell when an individual is a person strictly (i.e., how senile need one be in order no longer to be able to be a person strictly), and persons strictly might need to fear concerning their treatment (as well as the inadvertent mistreatment of other persons strictly) were such a practice not established. The social sense of person is a way of treating certain instances of human life in order to secure the life of persons strictly.

To recapitulate, we value children and our feelings of care for them, and we seek ways to make these commitments perdure. That is, social roles are ways in which we give an enduring fabric to our often inconstant passions. This is not to say that the social role person is merely a convention. To the contrary, it represents a fabric of ways of nurturing the high value we place on human life, especially the life that will come to be persons such as we. That fabric constitutes a practice of giving great value to instances of human biological life that can in some measure act as if they were persons, so that (1) the dignity of persons strictly is guarded against erosion during the various vicissitudes of health and disease, (2) virtues of care and attention to the dependent are nurtured, and (3) important social goals such as the successful rearing of children (and care of the aged) succeed. In the case of infants, one can add in passing a special consideration (4) that with luck they will become persons strictly, and that actions taken against infants could injure the persons they will eventually become.[20]

It should be stressed that the social sense of person is primarily a utilitarian construct. A person in this sense is not a person strictly, and hence not an unqualified object of respect. Rather, one treats certain instances of human life as persons for the good of those individuals who are persons strictly. As a consequence, exactly where one draws the line between persons in the social sense and merely human biological life is not crucial as long as the integrity of persons strictly is preserved. Thus there is a somewhat arbitrary quality about the distinction between fetuses and infants. One draws a line where the practice of treating human life as human personal life is practical and useful. Birth, including the production of a viable fetus through an abortion procedure, provides a somewhat natural line at which to begin to treat human biological life as human personal life. One might retort, Why not include fetuses as persons in a social sense? The answer is, Only if there are good reasons to do so in terms of utility. One would have to measure the utility of abortions for the convenience of women and families, for the prevention of the birth of infants with serious genetic diseases, and for the control of population growth against whatever increased goods would come from treating fetuses as persons. In addition, there would have to be consideration of the woman's right to choose freely concerning her body, and this would weigh heavily against any purely utilitarian considerations from restricting abortions. Early abortions would probably have to be allowed in any case in order to give respect due to the woman as a moral agent. But if these considerations are met, the exact point at which the line is drawn between a fetus and an infant is arbitrary in that utility considerations rarely produce absolute lines of demarcation. The best that one can say is that treating infants as persons in a social sense supports many central human values that abortion does not undermine, and that allowing at least early abortions acknowledges a woman's freedom to determine whether or not she wishes to be a mother.

One is thus left with at least two concepts of person. On the one hand, persons strictly can and usually do identify themselves as such—they are self-conscious, rational agents, respect for whom is part of valuing freedom, assigning blame and praise, and understanding obligation. That is, one's duty to respect persons strictly is the core of morality itself. The social concept of person is, on the other hand, more mediate, it turns on central values but is not the same as respect for the dignity of persons strictly. It allows us to value highly certain but not all instances of human biological life, without confusing that value with the dignity of persons strictly. That is, we can maintain the distinction between human biological and human personal life. We must recognize, though, that some human biological life is treated as human personal life even though it does not involve the existence of a person in the strict sense.

Conclusions

I wish to conclude now with a number of reflections reviewing the implications of distinguishing between human biological and human personal life, and between social and strict senses of person. First, it would seem that one can appreciate the general value of human biological life as just that. Human sperm, human ova, human cell cultures, human zygotes, embryos, and fetuses can have value, but they lack the dignity of persons. They are thus, all else being equal, open to socially justifiable experimentation in a way persons in either the strict or social sense should never be. That is, they can be used as means merely.

With infants, one finds human biological life already playing the social role of person. An element of this is the propriety of parents' controlling the destiny of their very young children insofar as this does not undermine the role *child*. That is, parents are given broad powers of control over their children as long as they do not abuse them, because very young children do in fact live in and through their families. Very young children are more in the possession of their families than in their own possession—they are not self-possessed, they are not yet moral agents. They do not yet belong to themselves. In fact, though persons strictly have both rights and duties, persons in the social sense are given moral rights but have no duties. Moreover, others must act in their behalf, since they are not self-determining entities. And when they act in their behalf, they need not do so in a manner that respects them as moral agents (i.e., there is no moral autonomy to respect), but in terms of what in general would be their best interests. Further, the duty to pursue those best interests can be defeated.

At least some puzzles about parental choice with regard to the treatment of their deformed infants or experimentation on their very young children can be resolved in these terms. Parents become the obvious ones to decide concerning the treatment of their very young children as long as that choice does not erode the care of children generally, or injure the persons strictly those children will become. And parents can properly refuse life-prolonging treatment for their deformed infants if such treatment would entail a substantial investment of their economic and psychological resources. They can be morally justified if they calculate expenses against the expected life-style of the child if treated, and the probability of success. Such a utility calculus is justified (i.e., it is in accord with general social interests in preserving the role child) insofar as it involves a sufficiently serious acknowledgment of the value of the role child (i.e., as long as such choices are not capricious and there is a substantial hardship involved so that such investment is "not worth it")[21] in order to maintain the practice of the social sense of person. Further, one may seek to justify social intervention in the form of legal injunctions to treat where such calculations by the parents are not convincing.

As to using very young children in experiments, they can be used in a fashion that adults may not, since they are not persons strictly. By that I mean it is morally permissible to consent on their behalf when the risk is minimal, the value pursued substantial, when such experiments cannot in fact be performed on adults, and when such treatment does not erode the use of the social sense of person. (One might picture here the trial of a rubella vaccine on children that was not intended to be of a direct benefit to those children, especially those who would grow to be misanthropic bachelors and thus never want to protect fetuses from damage.) Nor need one presume anything except that most small children who are vaccinated have in some fashion been coerced or co-opted into being vaccinated.

Consequently, with very young children one need not respect caprice in order to maintain the social sense of person. With free agents that is a different matter. Part of the freedom of self-determination is the latitude to act with caprice. For example, adults should be able, all else being equal, to refuse life-prolonging treatment; very young children should not. Surely difficult issues arise with older children and adolescents.[22] But the problems of dealing with free choice on the part of older children and adolescents attest to the validity of the rule rather than defeating it. With adults one is primarily concerned with the dignity of free agents, and what is problematic with respect to adolescents is that they are very much free agents.[23] In contrast, with small children one is concerned with their value (and the value of the social sense of person) and with not damaging the persons the children will become. In intermediate cases (i.e., older children) one must respect what freedom and self-possession do exist.

In summary, fetuses appear in no sense to be persons, children appear in some sense to be persons, normal adult humans show themselves to be persons. Is anything lost by these distinctions? I would argue not and that only clarity is gained. For those who hold some variety of homunculus theory of potentiality, it may appear that something is lost, for example, by saying that infants are not persons strictly. But how they could be such is, on the view I have advanced, at best a mystery. In this respect I would like to add a caveat lest in some fashion my distinction between persons strictly and persons socially be taken to imply that those humans who are only (!) persons socially are somehow set in jeopardy. It is one thing to say that it does not have great value. For example, the argument with regard to the social role *child* has been that a child is a person socially because it does indeed have great value and because the social sense of person has general value. Children receive the social sense of person because we value children, and moreover because the social sense of person has a general utility in protecting persons strictly. In short, there is no universal way of speaking of the sanctity of life; some life (personal life) has dignity, all life can have value, and human biological life that plays the social role person has a special value and is treated as human personal life.

What I have offered is, in short, an examination of the ways in which the biomedical sciences have caused the concept of person to be reexamined, and some of the conclusions of these examinations. These analyses lead us to speak not only of human biological versus human personal life, of strict versus social concepts of person, but to distinguish, with regard to the sanctity of life, the value of biological life, the dignity of strictly personal life, and the care due to human biological life that can assume the social role of a person.

Acknowledgments

An earlier version of this paper was read as a part of the Matchette Foundation Series, "The Expanding Universe of Modern Medicine," The Kennedy Institute and the Department of Philosophy, Georgetown University, Washington, DC, November 19, 1974. I wish to express my debt to George Agich, Thomas J. Bole, III, Edmund L. Erde, Laurence B. McCullough, and John Moskop for their discussion and criticism of the ancestral drafts of this paper.

Notes and References

[1]Black's Law Dictionary, 4th ed., rev., s.v. "death."

[2]For the first such statutory definition of death see: "Definition of Death," Kan. Stat. Ann., secs. 77–202 (1970).

[3]Willard Gaylin, "Harvesting the Dead," *Harper's Magazine,* **249** (September, 1974), 23–30.

[4]Ibid., p. 28.

[5]Ibid.

[6]Immanuel Kant, *Fundamental Principles of the Metaphysics of Morals, in Kant's Critique of Practical Reason and Other Works on the Theory of Ethics,* trans. Thomas K. Abbott, 6th ed. (1873; rpt. London: Longmans, Green and Co., 1909), p. 46; *Kants gesammelte Schriften,* 23 vols., Preussische Akademie der Wissenschaften, eds. (Berlin: Walter de Gruyter, 1902–56), IV, p. 428.

[7]Ibid.

[8]Immanuel Kant, *The Metaphysical Principles of Virtue: Part II of The Metaphysics of Morals,* trans. James Ellington (New York: Bobbs-Merrill, 1964), p. 23; *Akademie Textausgabe,* VI, 223.

[9]Black's Law Dictionary, 4th ed., rev., s.v. "person."

[10]Strictly, the current whole-brain-oriented definition of death distinguishes between a vegetative level of biological life (that which can exist after whole-brain death) and all higher levels. Report of the Ad Hoc Committee of the Harvard Medical School to Examine the Definition of Brain Death, "A Definition of Irreversible Coma," *Journal of the American Medical Association,* **205** (August 5, 1968), 85–88; Report of the Ad Hoc Committee of the American Electroencephalographic Society on EEG Criteria for Determination of Cerebral Death, "Cerebral Death and the Electroencephalogram," *Journal of the American Medical Association,* **209** (September 8, 1969), 1505–1510. *See also,* President's Commission for the Study of Ethical Problems in Medicine and Biomedical and Behavioral Research, *Defining Death* (Wahington DC: US Government Prinring Office, 1981). The commentary on the proposed Uniform Determination of Death Act is unsympathetic to regarding the brain's importance as lying in its sponsoring consciousness. For criticism of this position, *see* Richard M. Zaner (ed.), *Death: Beyond Whole-Brain Dead Criteria* (Dordrecht: Kluwer, 1988).

[11]I have treated these issues more fully elsewhere. H. Tristram Engelhardt Jr., *The Foundations of Bioethics* (New York: Oxford University Press, 1986), pp. 110–113.

[12]If one held that zygotes were persons (i.e., that persons begin at conception), one would have to account for how persons can split into two (i.e., monozygous twins), and for the fact that perhaps half of all persons die *in utero.* That is, there is evidence to indicate that perhaps up to 50 percent of all zygotes never implant. Arthur T. Hertig, "Human Trophoblast: Normal and Abnormal," *American Journal of Clinical Pathology,* **47** (March, 1967), 249–268.

[13]Even if such practices might involve some disvalue, it would surely not be that of taking the life of a person. Also, one must recognize that if intrauterine contraceptive devices act by preventing the implantation of the zygote, they would count as a form of abortion.

[14]Both Aristotle and St. Thomas held that human persons developed at some point after conception. See Aristotle, *Historia Animalium,* Book II, Chapter 3, 583 b, and St. Thomas Aquinas, *Summa Theologica,* Part 1, Q 118, art, 2, reply to obj. 2. See also St. Thomas Aquinas, *Opera Omnia,* XXVI (Paris: Vives, 1875), in *Aristoteles Stagiritae: Politicorum seu de Rebus Civilibus,* Book II, Lectio XII, p. 484, and *Opera Omnia,* XI, *Commentum in Quartum Librum Senteniarium Magistri Petri Lombardi,* Distinctio XXXI, Expositio Textus, p. 127.

[15]Michael Tooley, "A Defense of Abortion and Infanticide," *The Problem of Abortion,* ed. Joel Feinberg (Belmont, CA: Wadsworth Publishing Company, 1973), pp. 51–91.

[16]One might think that a counterexample exists in the case of sleeping persons. That is, a person while asleep is not self-conscious and rational, and would seem in the absence of a doctrine of potentiality not to be a person and to be therefore open to being used by others. A sleeping person is, though, a person in three senses in which a fetus or infant is not. First, in speaking of the sleeping person, one can know of whom one speaks in the sense of having previously known him before sleep. One therefore can know whose rights would be violated should that "person" be killed while asleep. His right to life would *in part* be analogous to a dead man's right to have a promise kept that had been made to him, when he was a self-conscious living person. In contrast, the fetus is not yet a person, an entity to whom, for example, promises can be anything but a metaphorical sense. Second, the sleeping man has a concrete presence in the world that is uniquely his, a fully intact functioning brain. Though asleep, the fully developed physical presence of the person continues. Third, the gap of sleep will be woven together by the life of the person involved: he goes to sleep expecting to awake and awakes to bring those past expectations into his present life. In short, one is not dealing with the potentiality of something to become a person, but with the potentiality of a person to resume his life after sleep.

[17]*See,* for example, R. C. Macmillan, H. T. Engelhardt, and S. F. Spicker (eds.), Euthanasia and the Newborn (Dordrecht: D. Reidel, 1987).

[18]It is important to note that whole-brain-dead adults fail to be persons in a social sense because they lack the ability for social interaction, not because they lack the potentiality to become persons. Markedly senile individuals can thus be persons socially long after they are no longer persons strictly.

[19]H. Tristram Engelhardt Jr., *The Foundations of Bioethics,* pp. 115–119.

[20]That is, once one is committed to refraining from killing infants because of a general interest in the value of the role *child,* one is committed to caring for infants so as not to injure the persons (strict) who will develop out of those infants. If one were to treat infants poorly, one would set into motion a chain of events that would injure the persons who would come to exist in the future (i.e., the persons such injured infants would become). But this presupposes that one has already decided on other grounds that infants should not be subject to infanticide. S. I. Benn fails

to make this point: *see* "Abortion, Infanticide, and Respect for Persons," in *The Problem of Abortion*, p. 102.

[21]It is not merely that it is difficult to impose a positive duty upon parents when that positive duty would involve great hardship, but that the actual object of that duty is not a person strictly. This gives considerable ground for skepticism regarding the current Baby Doe regulations. *See* McMillan et al. (eds.), *Euthanasia and the Newborn*.

[22]*See*, for example, John E. Schowalter *et al.*, "The Adolescent Patient's Decision to Die," *Pediatrics*, 51 (January 1973), 97–103: and Robert M. Veatch, ed., "Case Studies in Bioethics, Case No. 315," *Hastings Center Report*, 4 (September, 1974), 8–10.

[23]The issue with adolescents is not that they are not persons, but that special claims to act paternalistically can be made on their behalf by parents.

Once and Future Persons

W. R. Carter

> In cases like these it is pointless to insist on deciding in simple terms whether the statement is ' true or false'. Is it true or false that Belfast is north of London? That the galaxy is the shape of a fried egg? That Beethoven was a drunkard? That Wellington won the battle of Waterloo? There are various *degrees and dimensions* of success in making statements: the statements fit the facts always more or less loosely....[1]

So Austin says. Austin's thesis deserves attention when we run up against vexing questions as to whether certain "borderline" entities, a human fetus or infant, say, or an adult who has been irreversibly incapacitated but not killed in an automobile accident is a *person*. Can it happen in such cases that statements ascribing personhood "fit the facts" more or less loosely? It has been suggested —by David Lewis and Derek Parfit, among others—that there may be degrees of personhood.[2] Should that be so, there may be cases in which the "fit" between the facts of the situation and statements predicating personhood of some entity is "loose." The *fact* may be that the survivor of an accident is a person to some extent only. Should we think that a *statement* is true if and only if it fits, or corresponds to, the facts, and also think that the "fit" may be more or less loose, we will be well on the way to allowing that statements are true to the extent that they fit the facts. Regarded in this, all too hazy light, there may be a point to the claim that there is some truth, but some truth only, in statements saying of certain entities that they are persons.

The idea that there are gradations or degrees of truth poses tough problems in logic and semantics.[3] These problems are not addressed in this paper. It is the thesis that there are degrees of personhood that is the focus of attention.

I

It is often said that those of us who are persons have certain rights "inalienably," in the sense that these rights cannot be taken away from us by others.[4] Although there may be circumstances in which such rights may justifiably be overridden, or "infringed," there are no circumstances in which we can be stripped of such rights entirely. So it is claimed. There are, however, various arguments that can be used to challenge this thesis. One such argument proceeds along these lines:

A_1. Those of us who presently are persons existed, at some past time, as nonpersons,

A_2. (x) (x is a person iff x has a right to life),[5] so

A_3. Those of us who presently are persons are such that at some time in the past we existed without a right to life.

Granting (A_2), you can see to it that I no longer have a right to life by seeing to it that I no longer am a person. Assuming that I am not *essentially* a person—that is, assuming that the loss of the property of being a person is not necessarily the loss of *me*—others can see to it that I no longer have a right to life by seeing to it that I am no longer a person. If (A_1) is true, those of us who are persons are not essentially persons. That means that others can, at least theoretically, see to it that we exist as nonpersons. Granting (A_2), it follows that others *can* see to it that we exist without a right to life. So the right to life is not, on the assumptions in question, a right we have inalienably.

If the argument is sound, there was a time in the past at which we existed without a right to life. Had someone taken my life at this time they would not have been violating my right to life, since I had no such right at this time. But clearly I have such a right at present. Exactly when did I *acquire* this right?

On the assumption that (A_2) is true, we can answer that question by pinning down the time at which I became a person. Undeniably I am a person at present. Moreover, it is at least plausible to say of the physical organism that is located where I am located that *it* is a person. But since only one person is present in this location, it is plausible to say that I am identical with this physical organism. That means that we can replace the "When did I become a person?" question with "When did this organism become a person?" At some period in the past, this organism was an unborn fetus. But a fetus is not a person from the moment of its conception.[6] Unhappily, the prospects look dim for fixing a precise temporal point at which human organisms become persons.[7] But are the poor prospects a result of our failure to discern what is present, so to speak, in the nature of things or are they a reflection of the fact that there *is* no sharp "line" or boundry to be discovered? Perhaps there is a period during the organism's career during which the facts come to fit the "This is a person" statement (or a sequence of such statements) more and more closely. Perhaps the organism was a person to some extent, but to some extent only, during a relatively early part of its career. In that case, there is no sharp temporal line between my not being, and my being a person.

Of course granting (A$_2$) that means that as I came to be more and more a person I also came to gradually have more and more of a right to life. Can one have a so-called "basic" right only to a certain extent, or to some degree only? A. I. Melden is correct, I think, in saying that we do not ordinarily speak of rights being possessed by degrees.[8] But the fact that we do not speak this way hardly constitutes an argument for denying that we do, on occasion, have rights to some extent only.

II

Whatever we may think about the idea that basic rights can be more or less firmly possessed, we must reject at least one of the premises of Argument A if we think that such rights (e.g., the right to life) are possessed inalienably. Can (A$_1$) be successfully challenged? Consider Sydney Shoemaker's claim that:

> We tend to regard it as a necessary truth that persons, and only persons, are capable of thought, reason, and understanding.[9]

Even if this *is* necessarily true, it does not follow that those of us who are persons are essentially persons. Perhaps there are possible circumstances in which we exist without the capacities that Shoe-maker mentions. If that is so, it is possible for those of us who presently are persons to exist as nonpersons. There is reason to suspect that those of us who are people *did*, at one time, find ourselves in this position. There is only one way out for critics of (A_1), and that is to insist that such capacities are *essentially* possessed by those of us who have them. Taking myself as a "test" case, the argument looks like this:

B_1. I am essentially a being having the capacity to think and reason, i.e., there are no possible circumstances in which I exist without such capacities,

B_2. Necessarily I am a person if I have the capacity to think and reason, i.e., there are no possible circumstances in which I (or any other being) exists as a nonperson with such capacities or as a person without such capacities, so

B_3. I am essentially a person, i.e., there are no possible circumstances in which I exist as a nonperson.[10]

The argument is valid; so if the premisses are true, we have to accept:

C_1. $\Box$ $(\forall x)$ $(x = \text{Carter} \to \text{person } (x))$, which, conjoined with

C_2. $\Box$ $(\forall x)$ $(\text{person } (x) \to \text{right to life}(x))$, entails

C_3. $\Box$ $(\forall x)$ $(x = \text{Carter} \to \text{right to life } (x))$.

This means that there are no *possible* circumstances in which I (Carter) exist without the right to life. But in that case I never did, in actuality, exist without this right. So (A_1) must be false, if arguments B and C are sound.

But are they? I think that (B_1) and (C_1) are both false. At present I am sitting in a chair in my study. Located in the same chair is a physical organism—let's call it "Charley." On any remotely plausible capacity analysis of personhood, Charley is not *essentially* a person. It is all too possible that Charley exist in circumstances in which he (it) does not have the capacity to think or reason. Granting that much, it is hard to see how I, Carter, can be such that it is true both that I am identical with Charley and also have these capacities essentially. For any property P, if I have P essentially, I also have the further property of *having P essentially*.[11] It follows from the Indiscernibility of Identicals Principle that Charley would essentially have the capacity to think and reason were Charley identical with Carter and Carter essentially a thinking being. But Charley does not have the capacity to think essentially. So it is only if we are prepared to reject the claim that I am identical with Charley that we can accept (B_1).[12] The price of accepting (B_1) is that we deny that I am identical with Charley.

The trouble is that there is a strong, if not conclusive, argument for allowing that I am identical with Charley. The argument goes like this:

D_1. I presently am a person,

D_2. Charley presently is a person,

D_3. I presently am located in a chair in my office and Charley is located in this same chair,

D_4. There presently is one, and only one, person in this chair.

If these assumptions are correct, as I think they are, then Charley and I are one. For reasons just explained, that means that (B_1) is false. Thus advocates of Argument B must show that one of the premises of Argument D is false. (D_2) offers the likely point of attack. Those who reject this premise may say that although Charley is *related*, in some intimate though rather mysterious way, to a certain person, it still is not the case that *Charley* is (attributively) a person. It is I (Carter), and not Charley, who has the capacities that make something a person.

Anyone who endorses this picture of things will have to answer "Never" to the question "At what point in its development does a human fetus become a person?" They may say that as Charley gradually develops physiologically a different entity altogether (Carter) gradually acquires the capacities (thought, reason) that make their bearer a person. Alternatively, they may say that Charley's physiological development simply is *irrelevant* to Carter's capacity to think and reason. Neither of these claims is plausible. The problem in the last case is that there is a great deal of hard evidence clearly indicating that you can impair my psychological capacities by tampering with Charley's physiology. Nor, when we look closely, is it plausible to suppose that *Charley* does not acquire psychological capacities as he (it) develops physiologically. Situated as an unborn fetus, Charley *can* be hungry, say, or in pain. Supposing that I am essentially a person, and that a person is a being that can think and reason, *I* am not on the (immediate, anyway) scene when this happens. So it is not I, Carter, who is hungry or in pain. All of this strongly suggests that Charley *is*, even before he is born, a being having a certain range of psychological capacities. But once that is granted, there is no reason to deny that *Charley* eventually comes to have the full range of capacities that qualify their bearer as a person.

III

Thus (A_1) looks secure enough. It seems that critics of Argument A must focus their attention on (A_2). My instinct is to reject (A_2) on the grounds that not every person has a right to life. This move is not open to those who think—as Melden seems to—that every person necessarily has a right to life. [13] Anyone who thinks this can challenge (A_2) only by claiming that on occasion non-persons have a right to life. Does not a human fetus, say, have such a right even at times when it is not a genuine *person*?

I doubt it. But here we should guard against the tendency to focus attention exclusively on questions of possession of *rights*. Creatures lacking the attributes that are definitive of personhood often have some moral status or importance—are *morally considerable beings*, in the words of one commentator. [14] At the same time, persons surely count for more, morally, than do non-persons. Whatever its shortcomings, this position is not "speciesest," since its advocates need not analyze personhood as a *biological* classification. Of course it may be true, as Harry Frankfurt has argued, that the capacities that are definitive of personhood happen to be possessed only by members of the species *Homo sapiens*. [15] In cases in which members of this species lack certain capacities, and so fail to qualify as full-fledged persons, they seem to lack rights that persons possess. As Alan Donagan has persuasively argued, a man's insanity may give "those charged with looking after him the right of coercion...." [16] To the extent to which this happens, a man so situated may find himself lacking a right persons possess—"the right, subject to the moral law, to decide for himself what his own good is, and how to pursue it."[17] This does not mean that we have no duties towards someone so situated, or that he has no rights. It does mean that he fails to have rights others have and, to that extent, carries less moral weight than do many others.

There are, in short, important, if subtle, ties between *capacities* and rights. Since personhood is a matter of having certain capacities, there may well be reason for denying that nonpersons have a range of rights possessed by persons. Of course "moral weight" comparisons need not be based entirely upon *numbers* of rights. Part of what is involved in saying that one being has more moral weight than another may be that in cases of conflict, our obligations to the former override or outweigh our obligations to the latter. Something like this may be behind Robert Nozick's claim that although nonhuman animals "count for something," they have less "moral weight" than do people. Nozick asks:

> But *could* there be anything morally intermediate between persons and stones, something without such stringent limitations on its treatment, yet not be treated merely as an object? One would expect that by subtracting or diminishing some features of persons, we would get this intermediate sort of being.[18]

Perhaps those of us who presently are persons at one time had the status of "morally intermediate" beings, and as such carried less moral weight than we have as full-fledged persons! The argument, briefly stated, looks like this:

E_1. Nonpersons carry less moral weight than do persons,

E_2. Those of us who presently are persons existed as nonpersons, or persons only to some degree, at some time in the past, so

E_3. Those of us who are persons carried less moral weight at certain times in the past than we carry at present

Nozick suggests that (nonhuman) animals have the status of morally intermediate beings. If this argument is sound, those of us who are persons once were in much the same position. There is, however, a problem with this picture of things.

IV

At present, as a person, I carry a certain amount of "moral weight"; as a nonperson I carried less moral weight. But when, exactly, did I *become* a person? The question is crucial, since in acquiring the status of personhood we acquire a moral status we formerly lacked. It is tempting to reply to the question by saying that:

> Any attempt to draw a sharp line marking the onset of state of being a person is bound to be arbitrary....[19]

The trouble is that this reply meshes poorly with (E_1). Suppose—for purposes of illustration—that we acquire a right to life only upon acquiring the status of personhood. If that is so, and if it is somehow a conventional matter exactly when one becomes a person, *then it is a conventional matter exactly when one acquires a right to life*! This conflicts with the widely held, and I believe correct, view that possession of moral right is not a conventional matter that can somehow be stipulated or legislated by human fiat. Those of us who reject "conventionalist" accounts of moral rights must reject either the thesis that people generally carry more moral weight than non-people or the thesis that it is an "arbitrary" matter when people become people.

The question of whether moral judgments are somehow *conventional* is much too complex to tackle here.[20] I shall *assume* that in some important sense that is not the case. Anyone who doubts, or questions, this assumption is entitled to an argument that is not forthcoming in this paper. The assumption is that our moral obligations to other beings do not rest upon "arbitrary" or "conventional" decisions. Nor do such decisions underlie the fact that our obligation to one being sometime override our obligation to another being.

The problem with the "People Carry More Moral Weight" Thesis is that it seems to conflict with this anti-conventionalist sentiment. On any plausible analysis of personhood there will be tough "borderline" cases in which it is difficult to say whether something has *become* a person, or has *ceased to be a person*. There will, as one observer puts it, be a "penumbra region where our concept of a person is not so simple."[21] If the best that can be done in such cases is to draw an arbitrary line between what is, and is not, a person, the lines may be drawn differently in say, New York and Moscow. Assuming that people generally have a right to life and nonpeople do not, it then will turn out that whether certain entities have a right to life is something that depends on whether they happen to be located in New York or Moscow.

That is, I think, false. But how is the "result" to be avoided? We can reject the idea that people, and only people, have a right to life, but that is at best a stopgap measure. For if (E_1) is true the problem remains. Other things being equal, our obligation to save the life of a person overrides our obligation (if any) to save the life of a nonperson. Given the 'arbitrary line' approach to cases in the penumbra region, that implies that whether our obligation to save the life of certain entities is overridden by our obligation to save the life of others is something that depends on whether we happen to be in New York or Moscow.

V

It is here that the Degrees of Personhood Thesis may be relevant to the problem. The idea is basically that we:

> ...abandon the view that there is an abrupt transition to the status of a person and ...replace it by the view that being a person is a matter of degree. A one-year-old is much more of a person than a new-born baby or a fetus just before birth, but each of these is more of a person than the embryo.[22]

Those who take this seriously may say, paraphrasing Austin, that some ascriptions of personhood "fit the facts" more loosely than do others. As Charley gradually acquires certain capacities he becomes more of a person; the "fit" between the facts and statements ascribing personhood to Charley is accordingly less and less loose. There is then no *need* to accept the idea that the line between not being and being a person is arbitrary or conventional.

This tactic is, however, only as promising as the case for thinking that personhood really is a matter of degree. The case turns on the ideas (a) that being a person is a matter of having certain capacities, and (b) that the capacities in question may be possessed to a different extent by different entities, or by one entity at different stages of its career. Parfit, for one, seems to take both ideas seriously:

> Most of us now think that to be a person, as opposed to being a mere
> animal, is just to have certain more specific properties, such as ration-
> ality. These are matters of degree. So we might say that the fact of
> personhood is just the fact of having certain other peoperties, which are
> had to different degrees.[23]

Of course there is more to being a person than being rational. Perhaps,
as Frankfurt argues, the capacity for second-order volitions–the
capacity to want a certain desire to be our will should be added to the
list. Perhaps even then the list is incomplete. But even if it is, I agree
with Frankfurt's suggestion that any satisfactory analysis of personhood will rule out nonhuman animals, human fetuses and infants, and
even some adult humans as *persons*. It will, moreover, leave us with
borderline cases in which certain entities have the appropriate capacities to some extent only. If it is said of an entity so situated that it is
a person this statement fits the facts only loosely.

VI

Parfit also speaks of an "ancient principle" according to which the
welfare of people "is given absolute precedence over that of mere
animals."[24] How is this principle to be applied in cases where a certain
entity is a person to some extent, but to some extent only? I believe
that a plausible, if tentative, answer is this: In normal circumstances,
the welfare of x takes precedence over the welfare of y to the extent
to which x is more of a person than y. In such cases, our obligations
to insure x's welfare override our obligations to insure y's welfare to
the extent that x is more of a person than y.

The "normal circumstances" rider is meant to exclude cases in
which some special (and for our purposes, irrelevant) feature skewers
the "weight" of the relevant obligations. Here is a case that is abnormal. A boat containing a dog and a child is sinking. We can save either
the dog or the child but not both. To some extent the child is a person,
and so is more of a person than the dog. However, the dog can lead
us to a microbiologist who is lost and near starvation in a nearby forest.
The microbiologist has a cure for cancer. Only the dog can lead us to
him. Perhaps in these circumstances our obligation to save the dog
would override our obligation to save the child, even though the latter

is a person and the former is not. But since these are not normal circumstances, this does not pose a counter-example to the principle posed above.

The principle does not imply that marginal persons, or nonpersons, have no moral standing, or that we (people) have no obligations toward such beings. It does imply that in cases of unavoidable conflict our obligations to beings having more personhood generally outweigh or override our obligations to beings having less personhood. In conflict cases where the margin of personhood is negligible, there will be no reason (ignoring other factors) to suppose that our obligations to one party override our obligations to the other. In cases where the margin is large, and other things are equal, one obligation will heavily outweigh the other.

In some ways, it is a harsh view. (Certainly it will offend many people.) Frequently our emotional ties toward marginal people—say, the permanently incapacitated victim of an automobile accident—are much stronger than are our emotional ties to those who are full-fledged persons. In a perfect world we never would find ourselves in "conflict" situations in which we must choose between acting in the interest of a person and a marginal person. But the world is not perfect. In reality, it may turn out that our obligations conflict with our emotions.

VIII

There probably are no morally relevant features that enable us to justify the idea that human beings who are only marginally persons invariably count for more than do animals of other species. Granting that, one might be tempted to view the situation this way: Non-human animals count for no less, morally, than do *people*, since (1) the former frequently turn out to have the same (morally relevant) characteristics as humans who are only marginally persons, and (2) marginal (human) persons carry as much moral weight (have as much of a right to life, say) as do full-fledged persons. I suspect that it will be conceded by even the staunchest advocates of 'animal rights' that this view of things is a mistake. It is true that I have an obligation to promote the well-being of my dog, a being to which I have close emotional ties.

But if it comes down to choosing between the life of the dog and that of a person, my obligations to the latter override my obligations to the former. Still, (1) is correct. (Arguably many non-humans *are* marginal persons.) It is thesis (2) that must be rejected. The sad, but hard, fact is that our obligations to marginal persons are outweighed by our obligations to persons.

Notes and References

[1]"Truth," *Proceedings of the Aristotelian Society*, supplementary **24** (1950), 11–12.

[2]Cf. Parfit's "Later Selves and Moral Principles," *Philosophy and Personal Relations*, ed. by Alan Montefiore (Montreal, 1973) and Lewis' "Survival and Identity," *The Identities of Persons*, ed. by Amelie Oksenbedrg Rorty (California, 1976), p. 32.

[3]Many of these problems are addressed in David Sanford's interesting paper "Borderline Logic," *American Philosophical Quarterly*, **12** (1975), 29–39.

[4]A.I. Melden makes such a claim in *Rights and Persons* (California, 1977), p.167.

[5]Michael Tooley, "A Defense of Abortion and Infanticide," *The Problem of Abortion*, ed. by Joel Feinberg (Wadsworth, 1973), pp. 54–55.

[6]Judith Jarvis Thomson, "A Defense of Abortion,"*The Problem of Abortion, ibid.*, p. 122.

[7]*Ibid.*, p. 121.

[8]*Rights and Persons, op. cit.*, p. 188.

[9]*Self-Knowledge and Self-Identity* (Cornell, 1963), pp. 14–15.

[10]*Ibid.*, p. 15.

[11]Alvin Plantinga argues for this in *The Nature of Necessity* (Oxford, 1974), pp. 15–16.

[12]Cf. Fred Feldman, "Kripke on the Identity Theory," *The Journal of Philosophy*, **71** (1974), 665–676; cf. 667.

[13]*Rights and Persons, op. cit.*, pp. 167–170.

[14]Kenneth E. Goodpaster, "On Being Morally Considerable," *The Journal of Philosophy*, **75** (1978), 308–325.

[15]Harry G. Frankfurt, "Freedom of the Will and the Concept of a Person," *The Journal of Philosophy*, vol. **68** (1971), 5–20; cf. 10–11.

[16]Alan Donagan, *The Theory of Morality* (Chicago, 1977), p. 82.

[17]*Ibid.*, p.77.

[18]*Anarchy, State, and Utopia* (New York, 1974), pp. 35–40.

[19]Jonathan Glover, *Causing Death and Saving Lives* (Harmondsworth, 1977), p. 126.

[20]For a nice discussion of this issue *see* Gilbert Harman's *The Nature of Morality* (Oxford, 1977).

[21]Jane English, "Abortion and the Concept of a Person," *Canadian Journal of Philosophy* vol. **5** (1975), 233–243; cf. 235.

[22]*Causing Death and Saving Lives, op. cit.*, pp. 127–128.

[23]"Later Selves and Moral Principles," *Philosophy and Personal Relations*, ed. by Alan Montefiore (Montreal, 1973), p. 137

[24]*Ibid.*, p. 138.

Personhood
and the
Conception Event

Robert E. Joyce

This essay serves as a brief rationale for the claim that a *conceptus* or human zygote is essentially a human person. I will argue that the zygote is just as specifically and truly a person as you or I, though less developed. The idea is that conception or fertilization is the point at which a person—at least *one* individual person, possibly more—definitely begins to exist physically in the space-time world as we naturally and normally perceive this world. This is not offered as a probable conclusion, but as a reasonable certainty. The exact time at which a given conception event or fertilization process terminates may be quite uncertain. But I will maintain that there is definitely a moment of conception, a moment when the fertilization process is fundamentally complete and a single-celled zygote is essentially first in existence.

The basic format of my argument might be stated simply: every living individual being with the natural potential, as a whole, for knowing, willing, desiring, and relating to others in a self-reflective way is a person. But the human zygote is a living individual (or more than one such individual) with the natural potential, as a whole, to act in these ways. Therefore the human zygote is an actual person with great potential.

In the 1973 Supreme Court decision on abortion, Justice Harry Blackmun's majority opinion contained the claim that the unborn human individual is not a legal person in the whole sense.[1] Since that time the abortion issue has escalated into national and international prominence and does not seem likely to fade away soon. The positions of adherents to both sides of the underlying issues, concerning who is a person and when, are vigorously maintained. Closely associated groups of questions, especially relating to the value of the person as such, will probably assure increasing attention to the meaning of prenatal human life for generations to come.

As early as 1859 the American Medical Association Committee on Criminal Abortions issued a statement protesting the increasing practice of abortion and strongly asserting the importance of legal protection for prenatal human life.[2] As late as 1963, Planned Parenthood was claiming in a brochure that "an abortion kills the life of a baby after it has begun...birth control merely postpones the beginning of life."[3] In the 20th century, until the mid-60s, the vast weight of medical and legal opinion leaned toward the view that the life which started in the human being as the result of fertilization was worthy of serious moral and legal protection. What has happened in the last 10 years? Why have many jurists, scientist and other intellectuals attempted to deny essential personhood to the unborn or at least tried to redefine conception? What new scientific evidence or philosophical insight can justify this shift? I claim that there is none.

In fact, I will try to suggest that we need and can obtain greater philosophical understanding of the good common sense notion that persons can be tiny one-celled creatures just as wonderously as they can be complex trillion-celled creatures. First, I offer a definition of person and some comments by way of clarification. Second, a brief descriptive interpretation of the conception event is presented. Third, responses are given to some significant objections. And in conclusion, I mention a couple of major implications of the idea that the person exists at conception.

I. The Person

The first element of a sound interpretation of what occurs in human conception seems to be a definition of 'person.' 'Person' can be defined as a whole individual being which has the natural potential to know, love, desire, and relate to self and others in a self-reflective way. There are many alternate ways of phrasing the definition, depending upon different needs of emphasis. But it would seem to be crucial that we recognize a person as a natural being, and not simply as a functional being. A person is one that has the natural, but not necessarily the functional, ability to know and love in a trans-sensible or immaterial way. As soon as one would require a person to have the functional ability for this kind of activity, he or she would seem to be slipping into a subjectivistic elitism such that the comatose, senile, and retarded—even the sleeping—would not be regarded necessarily as persons. This is an unrealistic position that seems to be out of touch with the human condition. If nature has no essential value in our knowing and judging who or what is a person, independent of accepted functional abilities, then there is little hope for recognition of an objective nature transcending the limits of our personal consciousness in anything else.

Some philosophers have adopted what has been called the "developmentalist" interpretation of the beginning of a human person. Daniel Callahan views it this way. "(Abortion) is not the destruction of a human person—for at no stage of its development does the conceptus fulfill the definition of a person, which implies a developed capacity for reasoning, willing, desiring, and relating to others—but is the destruction of an important and valuable form of human life."[4] The language of the Supreme Court in *Roe v. Wade* is harmonious with this perspective. But I would suggest that a person is not an individual with a developed capacity for reasoning, willing, desiring, and relating to others. A person is an individual with a natural capacity for these activities and relationships, whether this natural capacity is ever developed or not. Individuals of a rational, volitional, self-conscious nature may never attain or may lose the functional capacity for fulfilling this nature to any appreciable extent. But this inability to fulfill their nature does not negate or destroy the nature itself, even though it may, for us, render that nature more difficult to appreciate and love. But that difficulty would seem to be a challenge for us as persons more than it is for them.[5]

Neither a human embryo nor a rabbit embryo has the functional capacity to think, will, desire, and self-consciously relate to others. The radical difference, even at the beginning of development, is that the human embryo actually has the natural capacity to act in these ways, whereas the rabbit embryo does not have and never will have it. For all its concern about potentialities, the developmentalist approach fails to see the actuality upon which these potentialities are based. Every potential is itself an actuality. A person's potential to walk across the street is an actuality that the tree beside him does not have. A woman's potential to give birth to a baby is an actuality that a man does not have. The potential of a human conceptus to think and talk is an actuality. This actual potential—not mere logical possibility—would seem to be a much more reasonable ground for affirming personhood than a kind of neo-angelic notion of personhood which requires actual performance of subjectively recognized spiritual activity.[6]

II. Conception: A Descriptive Interpretation

If the person is an individual entity with the natural, though not necessarily functioning, power to think, will, and relate self-reflectively, then when does such an individual actually begin to exist in the world of space and time? There would seem to be but one reasonable point at which to acknowledge the existence of a new individual person in this world. Conception is that moment when the so-called "fertilization" process is complete. From then on, a genetically and physically unique individual is present and growing. In the following description of the conception event, I wish to challenge or correct a few common misunderstandings about conception.

Before a sperm penetrates an ovum, these two cells are clearly individual cells and are parts of the bodies of the man and woman respectively. They are not whole-body cells as is the zygote cell which they crucially help to cause. They are body-part cells. The zygote is a single cell that is a whole body in itself. From within it comes all the rest of the individual, including the strictly intra-uterine functional organs of the placenta, amnion, and chorion, as well as the rest of the body that is naturally destined for extra-uterine life. The sperm and ovum are not potential life. They are potential *causes* of individual human life. They do not, even together, become a new human life. In the fertilization process, they become *causes* of the new human life.

Fertilization is a process. The process may take twenty minutes or several hours. But it has a definite conclusion. The moment at which this process terminates in the resulting zygote can be called the conception event. The sperm and ovum are specific, instrumental causes of the new human being. The man and woman are the main causes of the new human being. The man and woman are the main agents of this procreative effect. They cause an actual, not a potential, existence of a person in the space-time world. They do not cause a person to exist as a person, but they do cause (in an important, if partial, way) a person to exist in this world.

Parental bodily matter (the sperm and ovum) is a crucial element of procreative-causing on behalf of the new being, but is not the stuff out of which this unique bodily being is adequately constituted. The bodily matter of the zygote comes into existence by means of the bodily matter of the parents but does not come from their bodies. It only looks that way to the unphilosophical mind. The matter of the new person proper is constitutively different matter. The chromosomal uniqueness of the zygote is sufficient testimony to the radical difference of both form and matter in this new being. The unique matter of the zygote has traits similar to, but in no way identical with, those of the parents. With the perspective of an evolutionist, who once said that the evolution from nonlife to life was a "leap from zero to everything," we might say that the transition from parent body to offspring is a leap from zero self to all of self.

Moreover, the so-called fertilization process is not as passive as the terminology would suggest. The nuclei of the sperm and ovum dynamically interact. In so doing they both cease to be. One might say they die together. They really should not be said to unite. That suggests that they remain and form a larger whole. But the new single-celled individual is not an in tandem combo of the two parent sex cells. In their interaction and mutual causation of the new being, the sperm and ovum are self-sacrificial. Their nuclei are the subject of the fertilization process; the zygote is the result of this process. There is neither sperm nor ovum once the process of interaction is completed, even though cytoplasmic matter from the ovum remains. It is really a misleading figure of speech to say of the ovum that it is "fertilized" by the sperm, passively as a farmer's field is fertilized. It is proper rather to speak of the sperm–ovum interaction process. There is no such thing as a "fertilized ovum."

Obviously the new individual's growth is ever a process. But neither its coming into existence nor its final exit is a process. We

need to be paradoxical in our thinking, not simple-minded and re-
ductionistic, if we are going to appreciate both the process and the
non-process factors involved. In contemporary philosophy, when
process is valued on a par with substance the dignity of person and
nature are served and enhanced. But when process is enthroned
above substance, such that, in effect, the process itself is the only
substance, we are engaged in a self-deception fraught with episte-
mological and moral chaos.

At any given moment, a whole living substance—be it a peach
tree, a rabbit, or a human person—either is or is not alive. Once it is
alive, it is totally there as this particular actual being, even though it
is only partially there as a developed actuality. There is no such
thing as a potentially living organism. Every living thing is thoroughly actual, with more or less potential: actually itself; potentially
more or less expressive of itself. A one-celled person at conception
is an actual person with great potential for development and self-
expression. That single-celled individual is just as actually a person
as you or I, though the actual personhood and personality of the new
individual are, as yet, much less functionally expressed.[7]

In fact, the new personality is so little developed that we are not
yet able to recognize it functionally, unless we are willling to go beyond the vision of the eyeballs. Many are not willing. As one life
scientist remarked, in speaking of the users of the IUD: "...Ignor-
ance is bliss, for the blastocyst is only a little larger than the egg to
begin with, and its passage through the womb is unknown and unde-
tectable."[8] The issue thus becomes whether we are prepared to acknowledge the natural roots of the individual's personality within
this largely, though not entirely, undifferentiated stage. The genetic
differentiation of a zygote or a blastocyst, however, must be reasonably acknowledged as the natural roots of a *personality,* not of a
"dogality" or of a "rabbitality." The human zygote is a member of a
unique species of creature. It is not a genus, to which a species is
gradually attached. Such a process of attachment can occur in the
mind of the observer; but not in the reality of the observed.

No individual living body can "become" a person unless it already
is a person. No living being can become anything other than what it
already essentially is. From the perspective of the beginning of a
living thing, for such a being to become something essentially other
—say, for a "subpersonal human animal"[9] to *become* a person—it

would have to be a person before it was a person, so that it could be said to *become*.

Moreover, from the point of view of an adult, if, at this moment, I do not simply *have* a body, but in some radically natural sense I *am* my body, then, it is likewise perceptive to say or reasonable to conclude that I did not simply *get* a body at some point, but *was* that whole body naturally and radically at every point of *its* time as well as *its* space. Otherwise, I could never properly say such things as, "When I was conceived...."

III. Objections and Response

"The human *conceptus* is not necessarily an individual. But individuality is essential to personhood. Therefore, the *conceptus* cannot be reasonably regarded as a person." Proponents of this argument cite as evidence the fact of so-called "identical" twins and other multiple births resulting from the causality of a single ovum and sperm. They rightly insist that the living zygote which divides in half cannot be viewed as one identical human being dividing into two. But the evidence would seem to indicate *not* that there is *no* individual present at conception, but that there is at least one and possibly more. Jerome Lejeune of Paris, for instance, has indicated that individuality may be fully existent at the point of fertilization, but that thus far we do not have the technical capacity to discern how many individuals are present at that point.[10] Moreover, it seems to me that at this very early stage of human development there may occur at times a process of generation similar to that common in other species. In that case, we could say that one of the twins would be the parent of the other. The original zygote could be regarded as the parent of the second, even though we may never know which one was parthenogenically the parent.[11]

There is also the disputed evidence that in the first days of life, twins or triplets sometimes "recombine" into a single individual.[12] Actually, this could readily mean that one individual's body absorbs the body of the other, resulting in the latter's death at this particular, vulnerable stage in life.

The major type of objection to personhood at conception is some kind of developmentalism, such as gradual ensoulment. Develop-

mentalists claim to take into account life potential as well as life actual and thereby to give a more reasonable interpretation to the beginning of human personhood. But this approach fails on at least three
counts. It tends to confuse process in the collective with process in
the individual. It makes a typically utilitarian projection of mechanistic potential onto organic potential. And it seems to suffer from
the misleading, yet, popular, notion that man is a rational animal.

This gradualist approach does not distinguish sufficiently two
kinds of process. There is the process of the cosmos, as it might be
called within which living substances exist, and which causes these
individuals to exist in space and time. The individuals themselves
are not the subject of the process nor are they the cause of it. This
grand process of the whole of physical nature would seem to employ
individual substances, such as parents and their gametic cells, in the
causation of new individual substances. But there is another distinctive process, one that occurs within the living individual entities
themselves and one which they themselves cause. It is the process of
their own unique life and growth. This process is primarily caused
by the individuals; not by the environment and the whole process of
Nature. The individual in the womb of his or her mother is in charge
of the pregnancy, just as every individual in the womb of "Mother
Nature" is in charge of its own life and growth, even as it is thoroughly conditioned by its environment.

In their call for attention to potential life, gradualists have really
confused two different kinds of potency. The *potency to cause*
something to come into existence is improperly identifed with the
potency for this new being *to become* fully what it *is*. This latter kind
of potency applies only to living beings, since only these can grow or
become manifestly themselves. The zygote especially exemplifies
this later kind of potency—the *potency* of an existing being *to become* more expressly what it already is. The ovum and sperm particularly exemplify the first kind of potency—the *potency to cause*
something to come into existence.

One of the important sources of confusion regarding these radically different kinds of potency is the fact that they interweave and
interact. The potency to cause something to come into existence—
which is proper to the ovum, for instance—also entails the latent
function of disposing the newly caused being (the zygote) to become
fully what it is once it is. And once the zygote is, its potency to

become fully itself also entails the latent function of internally causing its own stages of organization and development. But the potency to cause something is radically different from the potency to become something.

The gametic and zygotic cells primarily illustrate this difference and this confluence of potencies. The ovum, for instance, besides having the potency to cause, together with the sperm, something else to come into existence, also has the potency to become fully itself once *it* is. And, as with all organic potencies, the potency is attained at the beginning of the ovum's existence (even though it is simply a body-part existence). The potency of an ovum to become fully itself, as an ovum, includes its capacity for containing 23 chromosomes, as well as its capacity for causing, together with a sperm cell, a new human being. Moreover, the new human individual, as a zygote, has its own radically different potency for becoming what it *is*, once it is. And within its potency for becoming what it *is*, is potency for causing embryonic, fetal, infant, child, adolescent, and adult stages of development, as well as the potency for causing new human beings through the instrumentality of its gametes.

In this age of the electron microscope, we now know that the matter of a zygote is essentially of the same structure as the matter of an adult. Even a hylomorphic theory, then, demands an acknowledgment that the zygote and the adult have the same formal cause.[13] Only the soul of a *person* could serve the zygote and embryo as an internal final cause of the development of a specifically human brain or of a human anything. No part of a person can be developed through the internal direction of a plant or an animal soul.[14]

A second major flaw in the gradualist approach is its subtle or not so subtle projection of a mechanistic model of development onto an organically developing reality. It fails to distinguish between natural process and artifactual process. Only artifacts, such as clocks and spaceships, come into existence part by part. Living beings come into existence all at once and then gradually unfold to themselves and to the world what they already, but only incipiently, *are*. Some developmentalists use the analogy of a blueprint in characterizing the zygote. But a blueprint never becomes part of a house, unless it is used to paper the walls.

Moreover, the human zygote is much more than a genetic package. It is a living being that has genes. We do not think that an adult is a package of organs, muscles, and bones, but that he or she is a being who *has* these structures. The whole of a living being is always, at every stage, much more than the sum of its parts.

A third major weakness in the gradualist approach is the implicit or explicit notion that a human person is a rational animal. But a person is not a rational animal any more than an animal is a sentient plant. Persons are animal-like, plant-like, rock-like, and God-like in many ways. We fall like rocks when dropped. We digest food like animals. And in our contemplative moments we act *like* God. Essentially, we are a wholly unique kind of *material* entity; even more different from animals than animals are from plants.

The latent idea that a human person is an "incarnate spirit" also seems to be at work in many who have a developmentalist approach. One's own body and biology are regarded as thoroughly subject to the superior and inevitably imperious judgments of mind and commands of will. The body is not valued as a vitally identifiable and intrinsic part of our person, but as an alien animal to be civilized by socialization and technology. By implication, then, one's own body is not regarded as an intrinsic revelation of person, but as a sophisticated instrument for personal use and eventual discard. The utilitarian society—well known for its tin cans, paper cups, disposable babies et al.—can find in the "developmental school" the heart of its rationale.[15]

In this view, nature is not a friend to be known and loved, but an alien, massive and impersonal monster ultimately to be outwitted and subdued. Thus the most immediately threatening and most symbolic part of this monster is one's body. One should not claim ownership of this body until one is sure he or she can handle it: until one is functionally capable of reasoning, desiring, and willing. These are the minimal criteria for a meaningful bodily existence, conferred by the person whose self-concept represents a refusal to be essentially (not exclusively) identified with body and biology. Such is the Cartesian legacy of the gradualist approach.[16]

Conclusions

The point in time when an individual person begins to exist in the spatio-temporal world is one of the most crucial metaphysical and social issues of our age. Induced abortion is a massive enterprise in both the East and West. One UN estimate cites a conservative figure of 50 million per year. Highly industrialized and presumably progressive nations, such as Japan and the USA, account for more than a million legal abortions in each nation per year. Philosophers on both sides of the Pacific are being challenged anew to make sense of human life beginnings.

In this country, the present movement for a Human Life Amendment, protecting all human life from conception until unneglectful natural death, underscores the need for distinctive philosophical contribution to the issue itself. In my view, prebirth individuals are now being dehumanized by definition through quasi-scientific and erroneous philosophical endeavors. The more subtle form of this dehumanization is the attempt to redefine conception in order to justify newer forms of chemical birth control that do not prevent ovulation but rather, systematically induce early abortion.[17] This sexual utilitarianism tends to obfuscate the metaphysical features of incipient personhood.

In order to put the issue of personhood and conception into its truest perspective, philosophers are being challenged to represent, clarify, and deepen our understanding of the value of the person in himself or herself. Because this value is, as it were, a seamless robe, our thinking must be woven from the natural, substantive, and non-functional levels of meaning. Otherwise, "quality of life" ethics becomes the "survival of the fittest," of the most functional; and the ethic itself becomes a non-ethic. I think we need an ethics sensitive to a deeper and richer vision of our dignity even as adults, who are dependently developing persons in the environment of space and time. Without appreciable insights into the inexhaustible process of personhood development, we will not be prepared to respect and protect the prenatal person.

Finally, there is hope that in the wisdom of both East and West we will come to realize how we can learn from the prebirth child concerning the meaning of human existence. Eventually, people could come to see, within the studies of fetology and ontology, many points of convergence and mutual resource.[18] And through cooperative endeavors in these and other disciplines, we could learn very much from our prenatal brothers and sisters as we ourselves continue to gestate within the premortal womb of space and time.

Notes and References

Notes and References

Notes and References

[1] US Supreme Court, *Roe v. Wade*, January 22, 1973, p. 47.

[2] American Medical Association Committee on Criminal Abortion, 1859. A resolution was adopted unanimously by the AMA, condemning abortion "at every period of gestation, except as necessary for preserving the life of either mother or child," and urging civil protection for fetal life.

[3] Planned Parenthood Federation of America, *Plan Your Children For Health and Happiness*, 1963.

[4] D. Callahan, *Abortion, Law, Choice and Morality* (New York, 1970), pp. 497–498.

[5] The recognition of a person involves, in part, a moral decision. This point is made effectively by John Noonan in the booklet, *How to Argue About Abortion* (New York, 1974), p. 10.

[6] Even the potential to receive actuation ("passive potency") is itself an actuality that is not had by something lacking it. There are subtle ways to overlook the actuality of potentiality in the case of personhood. E.g., Louis Dupré does so by using equivocally "person" and "personal" in his essay, "A New Approach to the Abortion Problem," *Theological Studies* (1973), 481–488.

[7] Cf. Robert and Mary Joyce, *Let Us Be Born* (Chicago, 1970), pp. 21–24.

[8] N. Berrill, *Person In the Womb* (New York, 1968), p. 32.

[9] An expression used by H. Tristram Engelhardt, Jr. in "The Ontology of Abortion," *Ethics* (April, 1974), 217, *et passim.*

[10] J. Lejeune, "The Beginning of a Human Being," a paper presented to the *Academie des Sciences Morales et Politiques,* Paris, Oct. 1, 1973. He is an eminent geneticist.

[11] This kind of explanation is likewise recognized in the position paper of Scientists For Life, "The Position of Modern Science on the Beginning of Human Life" (Fredericksburg, Virginia, 1975), p. 18.

[12] This phenomenon is alleged by Andre Hellegers, MD, in an article, "Fetal Development," *Theological Studies* (March, 1970), 4. Thomas Hilgers, MD, et al., dispute it. Cf. T. Hilgers, "Human Reproduction: Three Issues for the Moral Theologian," *International Review of Natural Family Planning,* I (1977), 115–116.

[13] Thomas Aquinas would be the first to admit it. His minimal conditions are underscored in places such as *Q.D. De Anima,* 2 ad 2; and 10 ad 1.

[14] For Acquinas, not only is the soul an internal final cause (as a formal cause) of the body; but it is also united with the body as an efficient cause. Cf. *Q.D. De Anima,* 9 and 10. In the light of contemporary genetic evidence, the entrancing and exiting of souls logically required for the mediate-animation interpretation would seem to degrade hylomorphic theory as an integrative theory of natural identity. Perhaps the most notable attempt at mediate animation today is that of J. Donceel. E.g., *see* his "Animation and Hominization," *Theological Studies,* (1970), 76–105. A thorough Thomistic critique of Donceel's argumentation is

given by W. Marshner, "Metaphysical Personhood and the I.U.D.," *The Wanderer,* (October 10, 1974), 7.

[15]An excellent critique of utilitarian ethics is given by G. Grisez, *Abortion: the Myths, the Realities, the Arguments* (New York, 1970), p. 317, *et passim.*

[16]Cf. D. Demarco, *Abortion In Perspective* (Cincinnati, 1974), *passim.*

[17]E.g., note the attempt of J. Diamond, MD, "Abortion, Animation, and Biological Hominization," *Theological Studies* (1975), 305–324.

[18]In fetology, A. M. Liley is perhaps foremost in the world. Cf., e.g., "The Fetus as a Personality," *The Australia-New Zealand Journal of Psychiatry,* **VI** (1972), 99–105. Or, e.g., J. Lejeune, "On the Nature of Man," *American Journal of Human Genetics* (March, 1970), 119–128.

Who Shall Count as a Human Being?

A Treacherous Question
in the Abortion Discussion

Sissela Bok

> "The temptation to introduce premature ultimates—Beauty in Aesthetics, the Mind and its faculties in Psychology, Life in Physiology, are representative examples–is especially great for believers in Abstract Entities. The objection to such Ultimates is that they bring an investigation to a dead end too suddenly."
> *I. A. Richards*, *Principles of Literary Criticism*, p. 40.

In discussions of abortion policy, the premature ultimate is 'humanity.' Does the fetus possess 'humanity'? How does one go about deciding whether a living being possesses it? And what rights go with such possession? These and similar questions have arisen beginning with the earliest speculations about human origins and characteristics. They are still thought central to the abortion debate. I propose to show in this paper that they cannot help us come to grips with the problem of abortion; indeed they obfuscate all discussion in this domain and lend themselves to dangerous interpretations precisely because of their obscurity.

The concept of 'humanity' is indispensable to two main arguments against abortion. The first defines the fetus as a human being and then concludes that abortion must be murder since it is generally considered murder to take the life of a human being. The second argument is designed to speak to those who do not believe that fetuses are human and cannot share, therefore, the conclusion that abortion is murder. It stresses, not the inherent wrong in individual acts of abortion, but rather the fearful consequences flowing from a social acceptance of abortion. According to this argument, it is impossible to draw a line in the period of prenatal development when humanity can be said to begin. There will therefore be no way to stop at early abortions, since they cannot be distinguished from later and yet later abortions; eventually society may even come to permit infanticide and the taking of lives generally. We are all at risk, according to such an argument, once we allow abortions to take place.

An analysis of these two arguments will show the ways in which the concept of 'humanity' operates as a premature ultimate.[1] I propose to substitute for this vague concept an inquiry into the commonly shared principles concerning the protection of life. These principles help to define workable rules for abortions and make it possible to draw a clear line between abortion, on the one hand, and the taking of life in infanticide, euthanasia, and genocide, on the other.

A. Humanity. A long tradition of religious and philosophical and legal thought has approached the problem of abortion by trying to determine whether there is human life before birth, and, if so, when it becomes human. If human life is present from conception on, according to this tradition, it must be protected as such from that moment. And if the embryo becomes human at some point during the pregnancy, then that is the point at which the protection should set in.

John Noonan[2] generalizes the predominant Catholic view as follows:

> "Once conceived, the being was recognized as a man because he had man's potential. The criterion for humanity, then, was simple and all embracing: If you are conceived by human parents, you are human."

Similarly, no less than ten resolutions had been introduced in Congress in the three months following the US Supreme Court's decisions on abortion.[3] These resolutions call for a constitutional amendment providing that

> "neither the United States, nor any state shall deprive any human being, *from the moment of conception,* of life without the due process of law..."

Others have held that the moment when implantation of the fertilized egg occurs, six–seven days after conception, is more significant from the point of view of individual humanity than conception itself. This view permits them to allow the intrauterine device and the 'morning after pill' as not taking human life, merely interfering with implantation. Whether or not one considers such distinctions to be theoretically possible, however, modern contraceptive developments are making them increasingly difficult to draw in practice.

Another widely shared approach to establishing humanity is that of stressing the time when the embryo first begins to look human. A photo of the first cell having divided in half clearly does not depict what most people mean when they use the expression 'human being.' Even the four-week embryo does not look human in this sense, whereas the six-week-old one begins to do so. Recent techniques of depicting the embryo and the fetus have remarkably increased our awareness of this early stage; this new 'seeing' of life before birth may come to increase the psychological recoil from aborting those who already look human—thus adding a psychological factor to the medical and other factors already influencing the trend to earlier and earlier abortions.

Another dividing line, once more having to do with perceiving the fetus, is held to occur when the mother can feel the fetus moving. Quickening—when these moments are first felt—has traditionally represented an important distinction: in some legal traditions, such as that of the common law, abortion was permitted before quickening, but considered a misdemeanor afterwards, until the more restrictive 19th century legislation was established. It is certain that the first-felt movements of the fetus represent an awe-inspiring change for the mother, comparable perhaps, in some primitive sense, to a 'coming to life' of the being she carries.

Yet another distinction occurs when the fetus is considered viable. According to this view, once the fetus is capable of living independently of its mother, it must be regarded as a human being and protected as such. The US Supreme Court decisions on abortion established viability as the "compelling" point for the state's "important and legitimate interest in potential life," while eschewing the question of when 'life' or 'human life' begins.[4]

A set of later distinctions cluster around the process of birth itself. This is the moment when life begins, according to certain religious traditions, and the point at which 'persons' are fully recognized in the law, according to the Supreme Court.[5] The first breaths taken by newborn babies have been invested with symbolic meaning since the earliest gropings toward understanding what it means to be alive and human. And the rituals of acceptance of babies or children have often defined humanity to the point where the baby could be killed if it were not named or declared accepted by the elders of the community or by the head of the household.

In the positions here examined, and in the abortion debate generally, a number of concepts are at times used as if they were interchangeable. 'Humanity,' 'human life,' 'life,' are such concepts, as are 'man,' 'person,' 'human being,' or 'human individual.' In particular, those who hold that humanity begins at conception or at implantation often have the tendency to say that at that time a human being or a person or a man exists as well, whereas others find it impossible to equate them.

Each of these terms can, in addition, be used in different senses which overlap but are not interchangeable. For instance, humanity and human life, in one sense, are possessed by every cell in our bodies. Many cells have the full genetic makeup required for asexual reproduction—so called cloning—of a human being. Yet clearly this is not the sense of those words intended when the protection of humanity or human life is advocated. Such protection would press the reverence for human life to the mad extreme of ruling out haircuts and considering mosquito bites murder.

It may be argued, however, that for most cells which have the potential of cloning to form a human being, extraordinarily complex measures would be required which are not as yet perfected beyond the animal stage. Is there, then, a difference, from the point of view of human potential, between these cells and egg cells or sperm cells? And is there still another difference in potential between the egg cell before and after conception? While there is a statistical difference in the likelihood of their developing into a human being, it does not seem possible to draw a clear line where humanity definitely begins.

The different views as to when humanity begins are little dependent upon factual information. Rather, these views are representative of different world-views, often of a religious nature involving deeply held commitments with moral consequences. There is no disagreement as to what we now know about life and its development before and after conception; differences arise only about the names and moral consequences we attach to the changes in this development and the distinctions we consider important. Just as there is no point at which Achilles can be pinpointed as catching up with the tortoise, though everyone knows he does, so everyone is aware of the distance traveled, in terms of humanity, from before conception to birth, though there is no one point at which humanity can be agreed upon as setting in. Our efforts to pinpoint and to define reflect the urgency with which we reach for abstract labels and absolute certainty in facts and in nature; and the resulting confusion and puzzlement are close to what Wittgenstein described, in *Philosophical Investigations,* as the "bewitchment of our intelligence by means of language."

Even if some see the fertilized egg as possessing humanity and as being "a man" in the words used by Noonan, however, it would be quite unthinkable to act upon all the consequences of such a view. It would be necessary to undertake a monumental struggle against all spontaneous abortions—known as miscarriages—often of severely malformed embryos expelled by the mother's body. This struggle would appear increasingly misguided as we learn more about how to preserve early prenatal life. Those who could not be saved would have to be buried in the same way as dead infants. Those who engaged in abortion would have to be prosecuted for murder. Extraordinary practical complexities would arise with respect to detection of early abortion, and to the question of whether the use of abortifacients in the first few days after conception should also count as murder. In view of these inconsistencies, it seems likely that this view of humanity, like so many others, has been adopted for limited purposes having to do with the prohibition of induced abortion, rather than from a real belief in the full human rights of the first few cells after conception.

A related reason why there are so many views and definitions is that they have been sought for such different purposes. I indicated above that many of the views about humanity developed in the abortion dispute seem to have been worked out for one such purpose: that of defending a preconceived position on abortion, with little concern for the other consequences flowing from that particular view. But there have been so many other efforts to define humanity and to arrive at the essence of what it means to be human—to distinguish men from angels and demons, plants and animals, witches and robots. The most powerful one has been the urge to know about the human species and to trace the biological or divine origins and essential characteristics of mankind. It is magnificently set forth beginning with the very earliest writings in philosophy and poetry; in fact, this consciousness of oneself and wonder at one's condition has often been thought one of the essential distinctions between humans and animals.

A separate purpose, both giving strength to and flowing from these efforts to describe and to understand humanity, has been that of seeking to define what a good human being is—to delineate human aspirations. What ought fully human beings to be like, and how should they differ from and grow beyond their immature, less perfect, sick or criminal fellow human? Who can teach such growth—St. Francis or Nietzsche, Buddha or Erasmus? And what kind of families and societies give support and provide models for growth?

Finally, definitions of humanity have been sought in order to try to set limits to the protection of life. At what level of developing humanity can and ought lives to receive protection? And who, among those many labelled less than human at different times in history—slaves, enemies in war, women, children, the retarded—should be denied such protection?

Of these three purposes for defining 'humanity,' the first is classificatory and descriptive in the first hand (though it gives rise to normative considerations). It has roots in religious and metaphysical thought, and has branched out into biological and archeological and anthropological research. But the latter two, so often confused with the first, are primarily normative or prescriptive. They seek to set norms or guidelines for who is fully human and who is at least minimally human—so human as to be entitled to the protection of life. For the sake of these normative purposes, definitions of 'humanity' established elsewhere have been sought in order to determine action—and all too often the action has been devastating for those excluded.

It is crucial to ask at this point why the descriptive and the normative definitions have been thought to coincide; why it has been taken for granted that the line between human and non-human or not yet-human is identical with that distinguishing those who may be killed from those who are to be protected.

One or both of two fundamental assumptions are made by those who base the protection of life upon the possession of "humanity." The first is that human beings are not only different from, but superior to all other living matter. This is the assumption which changes the definition of humanity into an evaluative one. It lies at the root of Western religious and social thought, from the Bible and the Aristotelian concept of the "ladder of nature" all the way to Teilhard de Chardin's view of mankind as close to the intended summit and consummation of the development of living beings.

The second assumption holds that the superiority of human beings somehow justifies their using what is nonhuman as they see fit, dominating it, even killing it when they wish to. St. Augustine, in *The City of God,*[6] expresses both of these anthropocentric assumptions when he holds that the injunction "Thou shalt not kill" does not apply to killing animals and plants, since having no faculty of reason,

> "therefore by the altogether righteous ordinance of the Creator both their life and death are a matter subordinate to our needs."

Neither of these assumptions is self-evident. And the results of acting upon them, upon the bidding to subdue the earth, to subordinate living matter to human needs, are no longer seen by all to be beneficial. The ancient certainties about man's preordained place in the universe are faltering. The supposition that only human beings have rights is no longer regarded as beyond question.[7]

Not only, therefore, can the line between human and non-human not be drawn empirically so as to permit normative conclusions: the very enterprise of basing normative conclusions on such distinctions can no longer be taken for granted. Despite these difficulties, many still try to employ definitions of 'humanity' to do just that. And herein lies by far the most important reason for abandoning such efforts: the monumental misuse of the concept of 'humanity' in so many practices of discrimination and atrocity throughout history. Slavery, witch-hunts, and wars have all been justified by their perpetrators on the ground that they thought their victims to be less than fully human. The insane and criminal have for long periods been deprived of the most basic necessities for similar reasons and excluded from society. A theologian, Dr. Joseph Fletcher, has even suggested as recently as last year that someone who has an IQ below 40 is "questionably a person" and that those below the 20-mark are not persons at all.[8] He adds that:

> "This has bearing, obviously, on decision-making in gynecology, obstetrics, and pediatrics, as well as in general surgery and medicine."

Here, a criterion for 'personhood' is taken as a guideline or action which could have sinister and far-reaching effects. Even when entered upon with the best of intentions, and in the most guarded manner, the enterprise of basing the protection of human life upon such criteria and definitions is dangerous. To question someone's humanity or personhood is a first step to mistreatment and killing.

We must abandon, therefore, this quest for a definition of humanity capable of showing us who has a right to live. We must seek, instead, common principles for the protection of life that reflect a clear understanding of the harm that comes from taking a life. Why do we hold life to be sacred? Why does it require protection beyond that given to anything else? The question seems unnecessary at first glance—surely most people share what has been called "the elemental sensation of vitality and the elemental fear of its extinction," and what Hume called "our horrors at annihilation."[9] Many think of this elemental sensation as incapable of further analysis. They view any attempt to say why we hold life sacred as an instrumentalist, utilitarian rocking of the boat which may loosen this fundamental respect for life. Yet a failure to scrutinize this respect, to ask what it protects and what it ought to protect, lies at the root not only of the confusion about abortion, but of the persistent vagueness and consequent abuse of the notion of the respect for life. The result is that everyone, including those who authorize or perform the most brutal killings in war, can protest their belief in life's sacredness. I shall therefore list the most important reasons that underlie the elemental sense of the sacredness of life. Having done so, these reasons can be considered as they apply or do not apply to the embryo and the fetus.

B. *Reasons for Protecting Life.*
1. Killing is viewed as the greatest of all dangers *for the victim*.
   –The knowledge that there is a threat to life causes intense anguish and apprehension.

  —The actual taking of life can cause great suffering.

  —The continued experience of life, once begun, is considered so
    valuable, so unique, so absorbing, that no one who has this
    experience should be unjustly deprived of it. And depriving
    someone of this experience means that all else of value to
    him will be lost.

2. Killing is brutalizing and criminalizing *for the killer.* It is a
   threat to others and destructive to the person engaging there-
   in.

3. Killing often causes *the family of the victim and others* to experience grief and loss. They may have been tied to the dead
   person by affection or economic dependence; they may have
   given of themselves in the relationship, so that its severance
   causes deep suffering.

4. All of society, as a result, has a stake in the protection of life.
   Permitting killing to take place sets patterns for victims,
   killers, and survivors that are threatening and ultimately
   harmful to all.

These are neutral principles governing the protection of life. They
are shared by most human beings reflecting upon the possibility of
dying at the hands of others. It is clear that these principles, if applied in the absence of the confusing terminology of 'humanity,'
would rule out the kinds of killing perpetrated by conquerors, witch-
hunters, slave-holders, and Nazis. Their victims feared death and
suffered; they grieved for their dead; and the societies permitting
such killing were brutalized and degraded.

Turning now to abortion once more, how do these principles apply to the taking of the lives of embryos and fetuses?

  C. *Reasons to Protect Life in the Prenatal Period.* Consider the
very earliest cell formations soon after conception. Clearly the reasons for protecting human life fail to apply here:

This group of cells cannot feel the anguish or pain connected with
death, nor can it fear death. Its experiencing of life has not yet begun; it is not yet conscious of the interruption of life, nor of the loss
of anything it has come to value in life, nor is it tied by bonds of
affection to others. If the abortion is desired by both parents, it will
cause no grief such as that which accompanies the death of a child.

Almost no human care and emotion and resources have been invested in it. Nor is such an early abortion brutalizing for the person voluntarily performing it, or a threat to other members of the society where it takes place.

Some may argue that one can conceive of other deaths with those factors absent, which nevertheless would be murder. Take the killing of a hermit in his sleep by someone who instantly commits suicide. Here there is no anxiety or fear of the killing on the part of the victim, no pain in dying, no mournings by family or friends (to whom the hermit has, in leaving them forever, already in a sense 'died'), no awareness by others that a wrong has been done; and the possible brutalization of the murderer has been made harmless to others through his suicide. Speculate further that the bodies are never found. Yet we would still call the act one of murder. The reason we would do so is inherent in the act itself and depends on the fact that his life was taken and that he was denied the chance to continue to experience it.

How does this deprivation differ from abortion in the first few days of pregnancy? I find that I cannot use words like 'deprive,' 'deny,' 'take away,' and 'harm' when it comes to the group of cells, whereas I have no difficulty in using them for the hermit. These words require, if not a person conscious of his loss, at least someone who at a prior time has developed enough to be or have been conscious thereof. Because there is no semblance of human form, no conscious life or capability to live independently, no knowledge of death, no sense of pain, one cannot use such words meaningfully to describe early abortion.

In addition, whereas it is possible to frame a rule permitting abortion which will cause no anxiety on the part of others covered by the rule—other embryos or fetuses—it is not possible to frame such a rule permitting the killing of hermits without threatening other hermits. All hermits would have to fear for their lives if there were a rule saying that hermits can be killed if they are alone and asleep and if the agent commits suicide.

The reasons, then, for the protection of lives are minimal in very early abortions. At the same time, many of these reasons are clearly present with respect to infanticide, most important among them the brutalization of those participating in the act and the resultant danger for all who are felt to be undesirable by their families or by others.

This is not to say that acts of infanticide have not taken place in our society; indeed, as late as the 19th century, newborns were frequently killed, either directly or by giving them into the care of institutions such as foundling hospitals, where the death rate could be as high as 90% in the first year of life.[10] A few primitive societies, at the edge of extinction, without other means to limit families, still practice infanticide. But I believe that the public acceptance of infanticide in all other societies is unthinkable, given the advent of modern methods of contraception and early abortion and of institutions to which parents can give their children, assured of their survival and of the high likelihood that they will be adopted and cared for by a family.

D. *Dividing Lines*. If, therefore, very early abortion does not violate these principles of protection for life, but infanticide does, we are confronted with a new kind of continuum in the place of that between less human and more human: that of the growth in strength, as the fetus develops during the prenatal period, of these principles, these reasons for protecting life. In this second continuum, it would be as difficult as in the first to draw a line based upon objective factors. Since most abortions can be performed earlier or later during pregnancy, it would be preferable to encourage early abortions rather than late ones and to draw a line before the second half of the pregnancy, permitting later abortions only on a clear showing of need. For our purposes, the two concepts of quickening and viability —so unsatisfactory in determining when humanity begins—can provide such limits.

Before quickening, the reasons to protect life are, as has been shown, negligible, perhaps absent altogether. During this period, therefore, abortion could be permitted upon request. Alternatively, the end of the first trimester could be employed as such a limit, as is the case in a number of countries.

Between quickening and viability, when the operation is a more difficult one medically and more traumatic for parents and medical personnel, it would not seem unreasonable to hold that special reasons justifying the abortion should be required in order to counterbalance this resistance: reasons not known earlier, such as the severe malformation of the fetus. After viability, finally, all abortions save the rare ones required to save the life of the mother,[11] should be prohibited, because the reasons to protect life may now be thought to be partially present; even though the viable fetus cannot fear death or suffer consciously therefrom, the effects on those participating in the event, and thus on society indirectly, could be serious. This is especially so because of the need, mentioned above, for a protection against infanticide. In the unlikely event, however, the mother should wish to be separated from the fetus at such a late stage,[12] the procedure ought to be delayed until it can be one of premature birth, not one of harming the fetus in an abortive process.

Medically, however, the definition of 'viability' is difficult. It varies from one fetus to another. At one stage in pregnancy, a certain number of babies, if born, will be viable. At a later stage, the percentage will be greater. Viability also depends greatly on the state of our knowledge concerning the support of life after birth and on the nature of the support itself. Support can be given much earlier in a modern hospital than in a rural village, or in a clinic geared to doing abortions only. It may some day even be the case that almost any human life will be considered viable before birth, once artificial wombs are perfected.

As technological progress pushes back the time when the fetus can be helped to survive independently of the mother, a question will arise as to whether the cut-off point marked by viability ought also to be pushed back. Should abortion then be prohibited much earlier than is now the case, because the medical meaning of 'viability' will have changed, or should we continue to rely on the conventional meaning of the word for the distinction between lawful and unlawful abortion?

In order to answer this question it is necessary to look once more at the reasons for which 'viability' was thought to be a good dividing-line in the first place. Is viability important because the baby can survive outside of the mother? Or because this chance of survival comes at a time in fetal development when the reasons to protect life have grown strong enough to prohibit abortion? At present, the two coincide, but in the future, they may come to diverge increasingly.

If the time comes when an embryo could be kept alive without its mother and thus be 'viable' in one sense of the word, the reasons for protecting life from the point of view of victims, agents, relatives, and society would still be absent; it seems right, therefore, to tie the obligatory protection of life to the present conventional definition of 'viability' and to set a socially agreed upon time in pregnancy after which abortion should be prohibited.

To sum up, the justifications a mother has for not wishing to give birth can operate up to a certain point in pregnancy; after that point, the reasons society has for protecting life become sufficiently weighty so as to prohibit late abortions and infanticide.

F. *The Slippery Slope.* Some argue, however, that such views of abortion could lead, if widely followed, to great dangers for society. This second major argument against abortion appears to set aside the question of when the fetus becomes human. It focuses, rather, on the risks for society—for the newborn, the handicapped, and the aged—which may stem from allowing abortions; it evokes the age-old fear of the slippery slope.[13] Because there are no sharp transitions in the period of fetal development, this argument holds, it would be unreasonable to permit abortion at one time in pregnancy and prohibit it shortly thereafter; in addition, it would be impossible to enforce such prohibitions. Later and later abortions may therefore be allowed, and there will be risks of slipping towards infanticide, euthanasia, even genocide.

The assumption made here is that once we admit reasons for justifying early abortions—reasons such as rape, incest, or maternal illness—nothing will prevent people from acting upon these very same reasons later in pregnancy or even after birth. If abortion is permissible at four weeks of pregnancy, then why not at four weeks and one day, four weeks and two days, and so on until birth and beyond? The reason that this argument possesses superficial plausibility has to do, once more with the concept of 'humanity.' Since all agree that the newborn infant is a human being, and since there are no ways of drawing clear lines before birth in the development of this human being, there appears to be no clear way of saying that the fetus is not human. On the assumption that humanity is the only criterion, there can then be a slippage from abortion to infanticide, with no clear dividing line between the two. Once more, then, 'humanity' turns out to be at stake. It is the concept providing the "slipperiness" to the slope—the dimension along which no distinctions can be made which make sense and are enforceable.

Once again, here, 'humanity' operates as a premature ultimate, bringing discussion to a dead end too soon. For the discontinuity which is not found in fetal development can be established by society, and indeed has been so established in modern societies permitting abortion. The argument that the reasons for aborting may still be declared to exist at childbirth completely ignores the reasons advanced against killing. These reasons grow in strength during pregnancy. Sympathy for the victim, grief on the part of those aware of the loss, recoil on the part of those who would do the killing, and a sense of social catastrophe would accompany the acceptance of infanticide by a contemporary democracy.

But, it may be asked, how can one know that these reasons would prevail? How can one be sure that the discontinuity will be respected by most, and that there will not be pressure to move closer and closer to an acceptance of infanticide?

The best way to answer such a question is to see whether that kind of development has actually taken place in one or more of the societies which permit abortion. To the best of my knowledge, the societies which have permitted abortion for considerable lengths of time have not experienced any tendency to infanticide. The infant mortality statistics of Sweden and Denmark, for example, are extremely low, and the protection and care given to all living children, including those born with special handicaps, is exemplary.[14] It is true that facts cannot satisfy those who want a logical demonstration that dangerous developments cannot under any circumstances come about. But the burden of proof rests upon them to show some evidence of such developments taking place before opposing a policy which will mean so much to women and their families, and also to show why it would not be possible to stop any such development after it begins to take place.

The fear of slipping from abortion towards infanticide, therefore, while superficially plausible, does not seem to be supported by the available evidence, so long as a cut-off time in pregnancy is established, either by law or in medical practice, after which all fetuses are protected against killing.[15]

I have sketched an approach to seeking community norms for abortion and tried to show the difficulties and dangers in using considerations of "humanity" to set such norms. Needless to say, individual choices for or against abortion will have to be more complex and influenced by religious and moral considerations.[16] Every effort must be made to show that abortion is a last resort. It presents difficulties not present in contraception, yet it is sometimes the only way out of a great dilemma.

Notes

[1]The focus in this paper is on abortion as a problem of social policy. Decisions made by individuals must take other factors into consideration. *See* S. Bok, "Ethical Problems of Abortion," *Hastings Studies*, January 1974 on which this article draws; first prepared for the Harvard Interfaculty Seminar on Children, chaired by Nathan B. Talbot.

[2]"An Almost Absolute Value in History," John Noonan, Jr., ed., *The Morality of Abortion*, p. 51, Harvard University Press, Cambridge, Massachusetts. For a thorough discussion of this and other views concerning the beginnings of human life, *see* Daniel Callahan, *Abortion: Law, Choice and Morality*, New York: Macmillan, 1970.

[3]"How the Constitution is Amended," *Family Planning/Population Reporter*, **2**, no. 3, p. 56.

[4]*Roe v. Wade, The United States Law Week* **41**, pp. 4227, 4229.

[5]*Ibid.*, p. 4227. For a discussion of this and other positions taken in the 1973 Supreme Court abortion decisions *see* L. Tribe, Foreword, *Harvard Law Review* **87**, 1–54, Nov. 1973.

[6]Augustine, *The City of God against the Pagans*, Book I, ch. XX, Cambridge: Harvard University Press, 1957.

[7]Christopher D. Stone, "Should Trees Have Standing? Toward Legal Rights for Natural Objects," *Southern California Law Review* **45**, 450–501, provides an interesting analysis of the extension of rights to those not previously considered persons, such as children, and a discussion of possible future extensions to natural objects.

[8]Joseph Fletcher, "Indicators of Humanhood: A Tentative Profile of Man," *The Hastings Center Report* **2**, no. 5, 1–4.

[9]Edward Shils, "The Sanctity of Life," in D.H. Labby, ed., p. 12, *Life or Death: Ethics and Options*, Seattle: University of Washington Press, 1968. David Hume, *Essay on Immortality*.

[10]William Langer, "Checks on Population Growth: 1750–1850," *Scientific American* **226**, no. 2, 1972.

[11]Every effort must be made by physicians and others to construe the Supreme Court's statement (*supra*) "If the state is interested in protecting fetal life after viability, it may go so far as to proscribe abortion during that period except when it is necessary to preserve the life or health of the mother" to concern, in effect, only the life or threat to life of the mother. *See* Alan Stone, "Abortion and the Supreme Court: What Now?" *Modern Medicine*, April 30, 1973, 33–37, for a discussion of this question and what it means for physicians.

[12]For an insightful discussion of this dilemma, *see* Judith Thomson, "A Defense of Abortion," *Philosophy and Public Policy* 1, no. 1, 47–66. My conclusions are set forth in detail in "Ethical Problems of Abortion" (*see* footnote 1).

[13]*See* S. Bok, "The Leading Edge of the Wedge," *The Hastings Center Report* 1, no. 3, pp. 9–11, 1971.

[14]Moreover, Nazi Germany, which is frequently cited as a warning of what is to come once abortion becomes lawful, had very strict laws prohibiting abortion. In 1943, Hitler's regime made the existing penalties for women having abortions and for those performing them even more severe by removing the limit on imprisonment and by including the possibility of hard labor for "especially serious cases." *See Reichsgezetzblatt*, 1926, Teil I, Nr. 28, par. 218, and 1943, Teil I, Art. I, "Angriffe auf Ehe, Familie, und Mutterschaft."

[15]Another type of line-drawing and slippery slope problem is that which would exist if abortions, once permissible, came to be coercively obtained in the case of mothers thought unable to bring up children, or in cases where deformed children were expected. To outlaw abortions out of a fear that involuntary abortions would take place, however, would be the wrong response to such a danger, just as outlawing voluntary divorces, operations, and adoptions on the grounds that they might lead to involuntary divorces, operations, and adoptions, would be. The battle against coercion must be fought at all times, with respect to many social options, but this is no reason to prohibit the options themselves.

[16]*See* "Ethical Problems of Abortion" (footnote 1).

On the Humanity
of the Fetus

Baruch Brody

In earlier essays[1], I have argued that if there is some point in the development of the fetus from which time on it is a living human being with all the rights of such an entity, then it would be wrong (except in one special case) to abort the fetus even to save the life of the mother, and then there should be strong laws prohibiting such abortions. In neither of those essays did I consider the question as to whether there is such a point, so I should like, in this essay, to outline an approach to finding the answer to that question.

I

Moral philosophers are in definite disagreement about this issue. Among positions as to when the fetus becomes a living human being with all the rights of such an entity are the claims that it does so at the moment of conception, at the time (around the seventh or eighth day) at which segmentation, if it is to take place, takes place at the time (around the end of the sixth week) at which fetal brain activity commences, at the moment of quickening (sometime between the 13th and 20th week) when the mother begins to feel the movements of the fetus, at the time (around the 24th week) at which the fetus becomes viable, i.e., has a reasonable chance of survival if born, and at the moment of birth.

It is difficult to see how one is to decide between these conflicting claims. The trouble is not merely that there are conflicting claims; the trouble is that the proponents of each of these claims seem to have somewhat persuasive arguments for their claims. Let us begin, therefore, by trying to understand each of these positions by looking at the arguments offered for them by their adherents.

The following seem to be the major reasons for supposing that the fetus becomes a living human being with all the rights that such an entity normally has at the moment of conception:

(1) At the moment of conception, the fetus is biologically determined by its genetic code. It is, from that point on, an individual unique creature, and everything that happens to it after that point is merely an unfolding of its unique selfhood. As Paul Ramsey[2] put it:

> Thus it might be said that in all essential respects the individual is whoever he is going to become from the moment of impregnation. He already is this while not knowing this or anything else. Therefore, his subsequent development cannot be described as becoming something he is not now. It can only be described as a process of achieving, a process of becoming the one he already is. Genetics teaches us that we were from the beginning what we essentially still are in every cell and in every generally human attribute and in every individual attribute.

(2) Until the moment of conception, the likelihood of whatever is present (the spermatozoa, the ova) developing into a clear-cut living human being is very small. But once conception has taken place, the resulting fertilized cell has a very high probability of developing into a clear-cut living human being. Indeed, four out of five of these entities do survive to birth. So these new entities, as opposed to the spermatozoa and ova, do have a human right to life. John Noonan, the leading advocate of this argument, put it[3] as follows:

> ...part of the business of a moralist is drawing lines. One evidence of the nonarbitrary character of the line drawn is the difference of probabilities on either side of it. If a spermatozoan is destroyed, one destroys a being which had a chance of far less than 1 in 200 million of developing into a reasoned being, possessed of the genetic code, a heart and other organs, and capable of pain. If a fetus is destroyed, one destroys a being already possessed of the genetic code, organs, and sensitivity to pain, and one which had an 80 percent chance of developing further into a baby outside the womb, who, in time, would reason.

(3) There is a continuity of development from the moment of conception on. There are constant changes in the fetal condition; the

fetus is constantly acquiring new structures and characteristics, but there is no one stage that is radically different from any other. Since that is so, there is no one stage in the process of fetal development, after the moment of conception, which could plausibly be picked out as the moment at which the fetus becomes a living human being. The moment of conception is, however, different in this respect. It marks the beginning of this continuous process of development and introduces something new which is radically discontinuous with what has come before it. Therefore, the moment of conception, and only it, is a plausible candidate for being that moment at which the fetus becomes a living human being. Roger Wertheimer (who is not himself an advocate of this position), summarized this argument[4] very well as follows:

> ...going back stage by stage from the infant to the zygote, one will not find any differences between successive stages significant enough to bear the enormous moral burden of allowing wholesale slaughter at the earlier stage while categorically denying that permission at the next state.

In order to understand the second position, the position that the fetus becomes a living human being at that moment at which, if it is to take place, segmentation takes place, it is necessary to remind ourselves of one or two key points about early fetal development. In a case in which there are identical twins, a primitive streak across the blastocyst signals the separation of the two twins. This occurs about the seventh day after fertilization. Although it occurs around the same time as implantation, it is an entirely separate process. And, of course, it is only a process that occurs when there are identical twins.

Now the argument for treating the fetus as a living human being only from this point on is really very simple. The individual in question only comes into existence, as a unique individual, at this point. Until then, for all we know, there may be two entities in the blastocyst. Paul Ramsey, the first one to raise this argument (which we shall label argument [4]), put it as follows[5]:

> It might be asserted that it is at the time of segmentation, not earlier, that life comes to be the individual human being it is thereafter to be...If there is a moment in the development of these nascent lives of ours subsequent to fertilization and prior to birth (or graduation from

college) at which it would be reasonable to believe that a human life
begins and therefore begins to be inviolate, that moment is arguably at
the stage when segmentation may or may not take place.

The next three positions to be considered (the moment at which
fetal brain activity begins, the moment of quickening, and the moment of viability), all share in common two basic ideas; the idea that
the fetus does not become human until some point at which it has far
more of the abilities and structures of a developed human being than
it has at the moment of conception, and the idea that there is some
point between conception and birth at which the fetus has acquired
enough of these significant characteristics so that it becomes plausible to think of the fetus as a living human being. These three positions differ only over the question as to when that point is.

The proponents of the claim that the fetus becomes a living human
being at about six weeks are primarily impressed with the fact that it
is about that time that electroencephalographic waves have been
noted,[6] and that, therefore, the fetal brain must clearly be functioning
after this date.   There are two main reasons for taking this
development to be the one that marks the time at which the fetus
becomes a living human being:

(5) It is just this indicator which is used in determining the moment of death, the moment at which the entity in question is no longer a living human being. So, on grounds of symmetry, it would seem
appropriate to treat it as the moment at which the entity in question
becomes a living human being.[7] Callahan (who is not entirely convinced by this argument) puts it as follows[8]:

> ...it is very rare, for instance, to find a discussion of when life begins
> (pertinent to abortion) related to a discussion of when life ends
> (pertinent to euthanasia and the artificial prolongation of life). Yet
> both problems turn on what is meant by human life, and the illumi-
> nation we gain in dealing with one of the problems will be useful
> when we deal with the other. Similarly, there is much to be said for
> trying to work out some consistent standards regarding the use of
> empirical data.

(6) One of the characteristics that are certainly essential to being a
living human being is that the entity in question is capable of conscious experience, at least of a primitive level. Before the sixth
week, it is not. Thereafter, it is. Consequently, that is the time at
which the fetus becomes a living human being.

Those who claim that the fetus becomes a living human being at the moment of quickening seem to be impressed with its significance for the following two reasons:

(7) Quickening is an indication of fetal movement. We would certainly want to think of the ability to move as one of those characteristics that are essential to living human beings (not, of course, only to living human human beings). So it is only at quickening, when there is a definite indication of fetal movement, that we are justified in thinking of the fetus as a living human being.

(8) There is an important sense in which it is true that quickening is that occasion from which the fetus can be perceived by other human beings. From that point on, the fetus can at least be felt by the mother. And anything which is not perceivable by other human beings cannot be thought of as a living human being only at the moment of quickening, only at the moment which it enters into the realm of the perceivable.

The following seems to be the main argument for supposing that the fetus becomes a living human being at the moment of viability:

(9) There is no doubt that the fetus is human from the moment of conception: it is certainly a human fetus. The question that we have to consider, however, is about when it becomes a living human being, and that is an entirely different matter. How can anything be a living human being if it is incapable of existing on its own? The fetus cannot do so until it becomes viable, so it cannot be a living human being until then. But when it does become viable, so it cannot be a living human being until then. But when it does become viable, then it has that degree of independence that is required for its being a living human being.

We come finally to the position that the fetus becomes a living human being only at its birth. It is clear that there is no special structure or capacity that it develops at that point. Indeed, this is so, in the last few months of pregnancy. So those who argue for birth as the moment at which the fetus becomes a living human being cannot be doing so on the grounds that it is then that they develop that which makes them a living human being. It is this that sets this position off from the last three that we have considered. What then are the main arguments for supposing that the fetus becomes a living human being at the moment of birth?

These seem to be the main points:

(10) As long as the fetus is in the mother, it is more appropriate to think of it as a part of the mother, rather than as an individual separate living human being. That status can accrue to the fetus, therefore, only when it emerges from the mother at the moment of birth.

(11) While it is true that the fetus has the capacity for independent existence from the time of viability, it does not actually have that independent existence until birth. Until then, its intake of oxygen and food, and its expelling of wastes (among other things) is parasitic on that of the mother's. So it is only at the moment of birth at which the fetus acquires that independence that is essential for its being a living human being.

(12) It is only after that birth that the fetus can interact with other humans and vice versa. But certainly it is this interaction, and not the mere abstract possibility of it, that is essential for being a living human being. So the fetus can be considered as a living human being only after its birth.

II

In light of the existence of these arguments, one can easily understand how many would conclude that something has gone wrong and that a fundamentally different approach is required. There is one such fundamental alternative that we should consider. It claims that there is a common but questionable presupposition of all the positions and arguments that we have been considering, viz., that there is an answer to the question as to when the fetus becomes a living human being with all the associated rights, an answer whose truth is independent of what we think or feel. It is just this presupposition that is rejected by the adherents of this fundamental alternative. According to them, the humanity of the fetus depends upon certain decisions involving it made by, and/or certain reactions to it of, other human beings.

One such version of this thesis is held by Professor John O'Connor.[9] While agreeing that once the criterion for humanity (for having the rights of a living human being) are settled, the question of the humanity of a fetus is a purely scientific one, O'Connor claims that it is we who have to decide upon the criterion of humanity, and, therefore, in an indirect way, it is we who ultimately have to make the decisions that will determine whether the fetus has the rights that a living human being has[10]:

> I suggest that the fundamental defect in Noonan's account is his assumption that the criterion of humanity needs to be discovered. Rather I suggest that we must decide what the criterion is to be.

In another passage, O'Connor puts his point as follows[11]:

> It is possible to agree with Noonan that, in a sense, it is certainly an objective matter whether or not a being is human, but point out that it becomes objective only when human beings have decided what the criterion of humanity is.

There are a variety of objections that might be raised against this view. But the most important one is that this seems to place the matter of human rights open to too many objectionable decisions. After all, there are all types of people with all types of prejudices about what is or is not required for being a living human being. And would we want to say that members of some minority group are really not living human beings just because they fail to meet the criterion of humanity established by some prejudiced majority, where the criterion in question reflects the prejudices of that majority group? If there were a vast majority prejudiced against redheads, and if that prejudice were reflected in some decision to include non-redheadedness as one of the necessary conditions for being a living human being, should we feel, as O'Connor's position seems to entail, that redheads then have no rights as living human beings? I think not. So O'Connor's position seems wrong.

O'Connor is, of course, aware of this possible objection, and he attempts to state the basis for his rejection of it in the following passage[12]:

> This is not, of course, to say that it is a subjective matter. Rather, there are good and bad reasons for deciding in the way we do.

There are two questions that naturally arise in response to such a remark: (1) what is the basis for distinguishing between good and bad reasons for such a decision, (2) is that basis such that it does not also serve as a basis for distinguishing between correct and incorrect answers to the question as to what is the criterion for being a living human being? It is extremely important that O'Connor should be able to say "no" in response to this second question, for if he cannot, then he will have avoided the subjectivism he recognizes as dangerous only at the cost of giving up his idea that the criteria of humanity are a matter to be decided upon and at the cost of adopting our position that there is an objectively true answer to the moral question "when is something a living human being?" It can, of course, be done. One way would be to provide a basis that picked out the reasons for several conflicting, but not all, proposed decisions as good reasons.

Let us now look at exactly what O'Connor does set out as his theory of good reasons and bad reasons. He puts it as follows[13]:

> The reason that humanity is of interest to a person concerned with the moral status of abortion is that he wants a way to decide the scope of the moral principle to the effect that the taking of human life is wrong. Hence humanity should be characterized in terms of those features which are in fact related to the moral sensibility of human beings...It would do little good to couch the criterion of humanity in such a way that the moral judgments we now make concerning human beings would be felt to have no moral force when applied to the 'newly qualified' humans, whose humanity was first recognized only by the new criterion of humanity.

There is something obscure about this passage in this context, because it looks like a statement of a basis for distinguishing good decisions from bad ones (although not, of course, correct answers from incorrect ones) rather than good reasons for adopting one decision from bad reasons for adopting one. So there seems to be a shift here in O'Connor's strategy. But it is close enough to what we want anyway, since having a criterion for good decisions is, for O'Connor's purpose, as satisfactory as having a criterion for good reasons for decisions.

But there is a great deal of trouble with O'Connor's answer. After all, it fails to meet the original objection. In a highly prejudiced society that has incorporated its prejudices into its criterion of humanity, it will be the case, according to O'Connor, that the only good criteria of humanity will be those that retain these prejudicial features. If we try to drop them out, so as to extend human rights to the minority in question, our new criterion of humanity will be a bad one, according to O'Connor's criterion, since it is not related to the moral sensibility of the human beings in that society. Going back once more to the case of redheads who, in their society, are not viewed as living human beings entitled to human rights, one would have to say that, according to O'Connor, it would be a bad decision to leave non-redheadedness out of the criterion for humanity.

Moreover, the reasons that O'Connor gives for adopting his criterion for good and bad decisions are not very convincing. To be sure, if one's decision extends the rules about human rights to entities not generally recognized as living human beings by most people, there is a very good chance that those rules will be broken in connection with these entities. In that sense, then, such a decision "would do little good." It might, however, do a great deal of good in other ways; it might serve an educational role (by making people think again about whether this minority is human), it might serve as an important protest of principle, etc. So it is not at all clear that such a decision would be a bad one. Moreover, the mere fact that a correct decision will not have the desired consequences should not, by itself, take away from its correctness.

In short, then, O'Connor's way of distinguishing good from bad decisions will not do, partially because it does not solve the problem raised by prejudiced societies and partially because the reasons for it are unconvincing. So we are left with our original objection to his whole approach, viz., that it seems to allow for the loss of human rights by some minority merely because some society has adopted a criterion of humanity that excludes that minority.

A very similar difficulty is faced by Professor Wertheimer, who offers another version of the thesis we are considering now. Professor Wertheimer is sympathetic to the suggestion[14] that:

> ...what our natural response is to a thing, how we naturally react, cognitively, affectively, and behaviorally, is partly definitive of how we ought to respond to that thing. Often only an actual confrontation will tell us what we need to know, and sometimes we may each respond differently, and thus have different understandings.

It is just this suggestion that we want to consider now.

Wertheimer himself is pessimistic of using his approach in the near future to resolve the problem of the status of the fetus. He feels

that there would have to be serious modifications in the fetal condition before we could have enough interactions with, and responses
to, the fetus so that we could see what we feel about it. Moreover,
what we would feel about this new type of creature is not clearly
relevant to the status of fetuses as they now exist.

It is not clear that Wertheimer's pessimism here is justified. Paul
Ramsey[15] has called our attention to one possibility of studying
human reactions to fetuses, even given their current condition, by
letting the response to viewing these pictures. Such an experiment
has never been carried out, but there is at least some data that suggests that its people see pictures of fetuses and of fetal behavior and
studying results might be quite significant. When Lennart Nilsson's
photographs of fetuses were published in *Life* in 1965, many readers
wrote in reporting their reactions, which generally tended to be that
they could no longer view the fetus, at least in the later stages, as a
disposable thing. Naturally, no conclusions should be drawn from
this small sample, but it does at least suggest that, were we to adopt
Wertheimer's approach to the question of when something is a living
human being, some progress could be made on the question of fetal
humanity.

Of course the question that we must consider first is whether we
want to adopt Wertheimer's approach. The question is, once more,
whether his approach can meet the problem raised by the responses
of a prejudiced society. He puts the problem as follows[16]:

> We surely want to say that Negroes are and always have been full-
> fledged human beings, no matter what certain segments of mankind
> may have thought, and no matter how numerous or unanimous those
> segments were.

His problem is how to reconcile this with the view that humanity is
determined by how people respond to the entity in question.

It is obvious that Wertheimer can meet this problem only by ruling out certain responses, only by showing that certain responses are
not relevant to the determination of the status of the creature in question. He offers us no full account of how he would do this, but we
get some idea of what he has in mind when we see how he rules out
as irrelevant to the status of the Negro slave the response to him of
the slaveholders[17]:

> We argue that his form of life is, so to speak, an accident of history, explicable by reference to special sociopsychological circumstances that are inessential to the nature of blacks and whites. The fact that Negroes can, and, special circumstances aside, naturally would be regarded and treated no differently than a Caucasian is at once a necessary and a sufficient condition for its being right to so regard and treat them.

Looking at this remark from our perspective, we see the following theory suggested here. A decision to count an entity as a human being is a good one only if it is in accord with the natural responses of clear-cut human beings to the entity in question. This criterion is thought reasonable because it reflects our privileged natural response and it solves the problem of the responses of the prejudiced society on the grounds that their responses are conditioned rather than natural.

Does this suggestion work? I think not. To begin with, can the distinction between the natural response and the sociopsychologically determined response do the work that Wertheimer needs it to do? I do not now want to raise the standard challenges to this distinction; rather, I would like to point out just how unclear the notion of the natural response is in this context. In one important sense, the slaveholder's response to the Negro is perfectly natural. Wertheimer himself points this out (without, I think, recognizing its full significance) in a footnote to his discussion of the slaveholder's response[18]:

> We develop our concept of a human through our relations with those near us and like us, and thus, at least initially, an isolated culture will generally perceive and describe foreigners as alien, strange, and not foursquare human.

On the other hand, there is an important sense in which the response of the integrationist is the natural one. As Huck Finn learnt, it is very natural, when put into situations in which one lives with Negroes, to respond to them as living human beings. So, the viability of Wertheimer's solution to the problem raised by the reactions of the prejudiced society is very unclear, because, in this context, appeals to what is a natural response seem to lead to hopelessly conflicting results.

Secondly, it is not clear that this criterion is reasonable. Why should we ascribe to the natural response the special status implicit in Wertheimer's proposal? Why should we suppose that the natural response gives us any deeper insight into the status of the entity in question than some historically and/or sociopsychologically determined response? I think that we have lurking here a new, and far more dangerous, naturalistic fallacy, the fallacy of supposing that what is natural is necessarily insightful.

Given, then, that neither Wertheimer nor O'Connor have been able to meet the problem of the prejudiced society, we cannot help but feel that this problem is going to destroy any version of this fundamental alternative approach that we are considering. This leads us then to the conclusion that we should return to the positions and arguments discussed in section I, that we should return to attempting to answer the question as to when the fetus really does become a living human being with all the rights that such an entity normally has.

III

As one looks over the various arguments, one is struck by the fact that they divide into two groups. The arguments in one group [consisting of arguments (3), (5), (6), (7), (8), (9), (11), and (12)] are based upon the consideration of the nature of fetal development, and whether there are some essential properties of being human that the fetus acquires at some point in its development that makes that point sharply discontinuous from what has come before it and which is therefore the time at which the fetus becomes a living human being. All of the arguments except (3) assert that there are such properties. We will be concerned with these arguments, and the theoretical issues that they raise, in the final sections of this article. But before doing so, we must first see why the remaining group of arguments [(1), (2), (4) and (10)], each of which is based upon some special feature, can be disregarded.

Argument (1), the genetic argument, begins from the biological fact that, at the moment of conception, the fertilized cell has the unique chromosomal structure that will be found in all of the cells of the living human being that will develop from this cell. But it goes it goes on to conclude from that that the entity in question is, from the moment of conception, whatever it is going to become, and that therefore it is a living human being. But how are these conclusions supposed to follow from the premise in question?

As we reflect upon the argument, it seems to come to the following: For any living human being there is a set of properties which can properly be considered the basic attributes of that human being. Some of these are shared with others, while others of these attributes are unique to the particular living human being in question. Now, that human being has these characteristics because of the chromosomal makeup of his cells; it is that that gives him these characteristics. But the fetus has already got that chromosomal makeup from the moment of conception; so it is also a living human being identical with the adult living human being.

The first thing that one should note about this argument is its somewhat dubious assumption that all of the basic characteristics of a human being (including those which he alone possesses) are genetically determined. That is to say, this argument presupposes that none of them is environmentally determined. It is, of course, difficult to decide whether or not this is so without an account of which characteristics are basic to that individual living human being, but the justification for assuming that they are all genetically determined is very unclear.

There is, moreover, a fallacy in the logic of this argument. Even if it is true that anything possessing the basic properties of some living human being *a* is identical with *a* and is like *a*, a living human being, and even if it is true that these characteristics are determined by the chromosomal structure that the fetus already has, it doesn't follow, and is clearly not true, that the fetus already has all of these characteristics; therefore, it does not follow that the fetus from which *a* will develop is identical with *a* and is like *a*, a living human being.

We turn now to argument (2), the argument from probabilities. This argument certainly does call attention to an important difference between fetuses (even from the moment of conception) on the one hand, and spermatoazoa and ova, on the other hand. But it is not immediately clear why this difference is supposed to show that the fetus (from the moment of conception) is a living human being.

There is one passage in which Noonan[19] explains the rationale for his argument:

I had supposed that the appeal to probabilities was the most common-
sensical of arguments, that to a greater or smaller degree all of us based our
actions on probabilities, and that in morals, as in law, prudence and negligence were often measured by the account one had taken of the probabilities. If the chance were 300,000,000 to 1 that the movement in the
bushes into which you shot was a man's, I doubt if many persons would hold
you careless in shooting; but if the chances were 4 out of 5 that the movement was a human being's few would acquit you of blame in shooting.

It is difficult to know what to make of this argument. To begin with,
Noonan switches, at the key point where he provides us with his
example, from the question of the morality of the action to the very
different question of whether you are to blame for something you
did, and the considerations that are relevant to one type of question
are not necessarily relevant to the other type. But more importantly,
the analogy is very unapt. In the case of the movement in the bushes,
the probability of 4/5 is the probability of the entity in question already being a living human being. In the case of the fetus, the probability of 4/5 is the probability of the entity in question developing
into a clear-cut living human being. And I cannot see how it follows
from the fact that one ought to suppose that the entity in question is a
living human being when the probability of its developing into a
clear-cut living human being is 4/5.

There is, however, a different suggestion that might be advanced
as to the relevance of this probability data. The following argument
is often advanced for the claim that the fetus, from the moment of
conception, is a living human being: from the moment of conception, the fetus has the potentiality for engaging in all of those activities that are typically human. Now, it is the potentiality for doing
that that makes one a living human being. So the fetus is a living
human being from the moment of conception. In connection with
this argument, the question naturally arises as to whether the fetus
does indeed already have these potentialities at the moment of conception. It might be felt that he does so just because of this difference in probabilities. An entity, like the fetus even from the moment
of conception, who has a 4/5 probability of developing into a being
that actually engages in these activities is an entity that has the potentiality of engaging in them.

Two additional points should be noted about this argument: (a)
the claim that the fetus has these potentialities can be reinforced by

an appeal to the facts about the genetic code. After all, there is, because of the presence of the genetic code in the fetal cells, a biological basis for these potentialities; (b) the argument we are considering now is not claiming that the fetus is a living human being because it is a potential human being. Such a claim is, of course, incoherent, for to be at a given time a potential P is precisely not to be, at that time, a P. Rather, it is arguing that the fetus is a living human being from the moment of conception because it has, from that moment, the potentiality of engaging in typically human activities, and it has that potentiality because of the probability consideration raised by Noonan (together, perhaps, with the facts about the genetic code).

There are several very difficult issues raised by this argument. The first has to do with what it is to possess the potentiality of engaging in human activities. It might be argued that, contrary to the argument we are considering, something more is required if the fetus is to be said to possess those potentialities. After all, consider as an example of a typical human activity, the activity of thinking. We have good reason to suppose that thinking can only take place in a living human being when certain physiological structures (in this case, neural structures) are present and operating. Would it not then be reasonable to say that an entity has the potentiality of engaging in these activities only when these structures are present? And if so, we certainly cannot say of the fetus, at the moment of conception, that it has the potentiality of engaging in thought. And, of course, a similar argument could be raised in connection with other typically human activities.

There is, moreover, a second objection to the whole argument that the fetus is a living human being because of its potentialities. Perhaps some actual human activity, in addition to the potentiality (in any sense) of human activity, is required if the entity in question is to be a living human being? There is, after all, a certain intuitive plausibility to this claim. If we suppose that a sufficient condition for being a living human being is the engagement in typically human activities, then isn't it plausible to suppose (admittedly, it does not follow) that the potentiality of engaging in these activities is sufficient only for being a potential human being? So even if we grant the fetus's potentialities on the basis of Noonan's statistical considerations, it hardly seems to follow that the fetus is a living human being.

We come now to argument (4), the argument for the time of segmentation as the time that the fetus becomes a living human being.
If we were to formalize this argument, it would run as follows:

(a) Until the time of segmentation, but not thereafter, it is physically possible that more than one living human being develop out of
that which resulted from the fertilization of the ovum by the sperm.

(b) Therefore, that which results from the fertilization of the ovum
by the sperm (i) cannot be a living human being until that time of
segmentation and (ii) is a living human being after that time.

The trouble here is, of course, that it is totally unclear as to how
either part of (b) follows from (a). Why should (i) be true just because (a) is true? The following suggests itself: if the fetus were a
living human being at some time before segmentation, and then it
were to split into two living human beings, then we would have one
living human being becoming two, and that is not possible. Unfortunately, although initially persuasive, this argument must be rejected. One ameba can become two, [20] so why can't one living human being become two?

It is equally unclear as to why it should be thought that (ii) should
follow from (a). Even if we suppose that (i) follows, this only means
that the fetus, by the time of segmentation, has passed one hurdle in
its path toward becoming a living human being. It is   now a unique
individual that will not split into two others. But it certainly does not
follow that it has passed all the hurdles; it certainly does not follow
that it is now a living human being.

We come finally to the very weak argument (10), the argument
that claims that the fetus is not a living human being until the moment of birth because, until then, it is only a part of the mother since
it is in the mother. The inference here seems to be from the fetus (a)
being in the mother (b) to its being a part of the mother and not an
independent entity. But one certainly cannot infer merely from the
fact that a is in b that a is a part of b (and certainly not that a is not an
independent entity). I am in this room, but I am certainly not a part
of this room. Jonah in the whale was not a part of the whale. To be
sure, the fetus is in the mother in the stronger sense that it is dependent upon the mother, and so on. That is, of course, a very different
matter, and argument (11) which raises it is a serious argument
which cannot be dismissed so easily and which we will consider below. For the moment, we need only note that the mere fact that the

fetus is in the mother says nothing about its status as a living human being.

IV

We turn finally to a consideration of the remaining arguments. With the exception of (3), the argument from the continuity of fetal development, they all seem to be of the following form:

(a) there is a property P which is such that every living human being must have it; it is essential for its being a living human being.

(b) when the fetus acquires P, it becomes a living human being. And even argument (3) is best viewed as denying that, at any point after conception, the fetus acquires any property that satisifies premise (a).

In this final section, I would like to consider some of the difficulties that might be involved in trying to defend an argument of this type and briefly sketch a possible way of dealing with them. The first point to be noted is that none of the arguments, as currently formulated, is valid. After all, if there are two such properties P and Q, the mere possession of one cannot make the fetus a living human being so long as it does not possess the other. So (b) cannot follow from (a). In order to meet this point, one would have to modify our arguments so that they would be of the following structure:

(c) there is a property P which is such that every living human being must have it; it is essential for its being a living human being.

(d) by the time an entity acquires P, it has every other property Q which is essential for its being a living human being.

(e) when the fetus acquires P, it becomes a living human being.

Even with this reformulation, however, none of these arguments would be valid. As they stand, all that the premises guarantee is that, by the time the fetus has acquired P, it will have also acquired all those other properties that living human beings must have, i.e., it will have satisfied all of the necessary conditions for being a living human being. But they do not guarantee that the fetus will have satisfied any condition that is sufficient for its being a living human being; i.e., they do not guarantee that (c) is true.

So the first major problem that any proponents of our arguments will have to consider is the following: is it possible to develop a theory of what properties are essential for being a human being according to which it will be true that

(d) when an entity acquires every property which is essential for its being a living human being, it becomes a living human being? It is only if such a premise is true that any of the arguments that we are considering will be valid.

There is, of course, a second problem that the proponents of any of our arguments will have to face: how are we to determine whether or not the possession of a given property is or is not essential for being a living human being? It should be kept in mind that this is an extremely difficult problem given our reformulation of the arguments. It is not enough that we know whether or not the given property P is essential for being a living human being. It would seem that we have to have a complete list of all of the properties that are essential for being a living human being. Otherwise, how would we know that the premise of type (d) is true? And how are we going to ascertain what is the full list of these properties essential to being human?

So much for the problems. Now for a brief sketch of a possible way of dealing with them.[21] Let us begin with the following account[22] of when it is that an object has a property essentially. An object *a* has a property P essentially just in the case that *a* cannot lose it without going out of existence, while *a* has P accidentally just in the case that *a* can change and lose the property without going out of existence. Thus, it is an accidental property of my tree that it has 832 leaves on it since it could grow an additional leaf and still continue to exist. But it is an essential property of it that it is a tree, for if it were chopped down and cut into lumber (so that it were no longer a tree), it would no longer exist. Now, we shall say that any property had essentially by some object and accidentally by none (whether actual or potential) determines a natural kind, and that the set of objects having that property is a natural kind. In short, a natural kind is a set of objects each of which has a certain property essentially, and nothing else has that property. Now, the set of white objects is not a natural kind, since the only property that all white objects, but nothing else, has is the property of being white, and not all white objects have that property essentially. After all, my desk, which is white, could be painted blue and still exist. On the other hand, the set of living human beings seems to be a natural kind, since there seems to be a property which only its members have (the property of being a living human being) and they all seem to have it essentially. After all, when we die, when we stop being a living human being, we (but not our body and perhaps not our soul) cease to exist. Surely that is why we treat death so differently than anything else that might happen to us. As Wittgenstein[23] put it, "Death is not an event in life: we do not live to experience death."

With these preliminaries out of the way, I should now like to introduce several claims about what the essential properties are for membership in a natural kind:

(1) Only the possession of properties had essentially by every member of a natural kind is essential for membership in that natural kind.

(2) Only the possession of all properties had essentially by every member of a natural kind is sufficient for membership in that natural kind.

Claim (1) tells us what properties are such that their possession is essential (necessary) for membership in natural kinds whereas claim (2) tells us what properties are such that their joint possession is sufficient for membership in the class in question.

Both of these claims are false if we are dealing with classes in general, and not just with natural kinds. Consider once more the class of white objects which is, as we have seen, not a natural kind. Certainly, being white is essential for membership in that class but being white is not, as we have seen, an essential property of every member of that class. So claim (1) would be false for this nonnatural kind. Similarly, the only properties had essentially by every member of this class are those had essentially by all colored objects (i.e., those had essentially by all objects and the property of being colored), but the possession of those properties is not sufficient for membership in the class of white objects (a blue object would, after all, also have them). What is sufficient is being white. So claim (2) would also be false for this nonnatural kind.

Intuitively, what is happening here is the following: assume that, for every class, there are some properties[24] which are such that the possession of each of them is necessary and their joint possession is sufficient for membership in the class in question. Now (1) claims that the only necessary properties are those had essentially by all members of that class and (2) claims that their joint possession is the only sufficient condition. This must be false in the case of nonnatural kinds for precisely what they lack are some properties that their members, and only their members, have essentially and the possession of which could therefore be the necessary and sufficient conditions for membership in the class in question. But in the case of natural kind, where all members have some properties essentially, and nothing else has them at all, it is plausible to conjecture that it is the possession of these properties, and only them, which is necessary and sufficient for membership in the natural kind. This is, of course, precisely what is claimed in (1) and (2).

I have no proof that (1) and (2) are true. But they are intuitively plausible and no counter-examples seem to be immediately forthcoming. So I will tentatively adopt them in this sketch. Given, then, (1) and (2), and our previous claim that humanity is a natural kind, we are now in a position to properly evaluate our remaining arguments. Their common structure was the following:

(c) there is a property P which is such that its possession is essential for being a living human being.

(d) by the time an entity acquires P, it has every other property Q which is essential for being a living human being.

(e) when the fetus acquires P, it becomes a living human being.

And there were two major problems with each of these arguments, viz., how to tell whether their essentialist claims are true and how to fix up their logic so that (e) will follow from the appropriate (c) and (d). We can now see how to solve these problems. Since humanity is a natural kind, given assumption (1), the only properties essential for being a human being are those had essentially by every human being, i.e., those which are such that their loss would mean that the entity in question has gone out of existence. We can therefore use the going-out-of-existence test to determine the truth of claims (c) and (d) in any given argument. And given assumption (2), (e) does follow straightforwardly from (c) and (d).

In short, then, our technical excursus has put us into a position for dealing with the problem of the essence of humanity. And it suggests to us that the soundest argument is (5), the brain-function argument. After all, it and only it seems to rest upon the claim that the fetus becomes a living human being when it acquires that characteristic which is such that its loss entails that a living human being no longer exists. But this, like all other points in this sketch, needs further investigation.

V

Where, then, do we stand on this vexing issue of fetal humanity? We have seen pretty clearly that it is not a matter to be resolved upon the basis of human decisions and reactions: it is, in that sense, a more objective matter. We have seen, moreover, that the crucial objective factors are ones having to do with the essence of humanity and not ones having to do with genetic codes and with probabilities of development. Finally, we have sketched (although certainly not proved the truth of) an approach to the essence of humanity according to which the fetus would be a living human being from about six weeks, the time at which we begin to note fetal brain activity.

Notes

[1]"Abortion and the Law," *Journal of Philosophy* (1971) and "Abortion and the Sanctity of Human Life," *American Philosophical Quarterly* (1973).

[2]"Points in Deciding about Abortion" in Noonan's *The Morality of Abortion* (Harvard, 1970), pp. 66–67. In that essay, Ramsey seems to alternate between that position and the time-of-segmentation position.

[3]In "An Almost Absolute Value in History" in his *The Morality of Abortion* (Harvard, 1970), pp. 56–57.

[4]In his "Understanding the Aboriton Argument," *Philosophy and Public Affairs* (1971), p. 83.

[5]*Op. cit.*, p. 66.

[6]It is interesting to note that Glanville-Williams, in his *The Sanctity of Life and the Criminal Law* (Stevens, 1961) thinks of the presence of fetal brain ac-tivity as a good compromise date for the beginning of fetal humanity, but only because he mistakenly thought that fetal brain activity is first detectable in the seventh month.

[7]This argument is sometimes turned around in very strange ways. Thus, writing in a letter to the New York Times (dated March 6, 1972), Cyril C. Means, Professor of Constitutional Law at New York Law School, argued as follows:

> An adult heart donor, suffering from irreversible brain damage, is also a living human "being," but he is no longer a human "person." That is why his life may be ended by the excision of his heart for the benefit of another, the donee, who is still a human person. If there can be human "beings" who are nonpersons at one end of the life span, why not also at the other end?

Professor Means seems to be missing the point. If we took the analogy he, and the argument we are considering both suggest, then, in his termiology, the fetus will be a person, as well as a human being, from the sixth week on and will, from that point on, be entitled to the full rights of such an entity.

[8]In his *Abortion: Law, Choice, and Morality* (Macmillan, 1970), p. 334.

[9]In his "On Humanity and Abortion," *Natural Law Forum* (1968).

[10]*Ibid.*, p. 13.

[11]*Ibid.*

[12]*Ibid.*

[13]*Ibid.*, p. 130.

[14]*Op. cit.*, p. 92.

[15]*Op. cit.*, 74. It should be noted that Ramsey himself is dubious about the value and importance of these results: "Medical science knows the babies to be present in all essential respects earlier in fetal development than the women who wrote into *Life* magazine perceived them in the pictures. It is the rational account of the nature of fetal development that matters most."

[16]*Op. cit.*, p. 86.

[17]*Ibid.*, p. 87.

[18]*Ibid.*

[19]"Deciding Who Is Human," *Natural Law Forum* (1968), p. 136.

[20]A whole literature has arisen about this so-called splitting problem and the difficulties that arise for the theory of personal identity. See, for example, D. Parfit's "Personal Identity," *Philosophical Review* (1971). But the example of the amoeba shows that, one way or another, we can live with it.

[21]I hope to be able to present in the not-too-far future a fuller version of this sketched approach.

[22]For a full defense of this account, see my "De Re and De Dicto Interpretations of Modal Logic," *Philosophia* (1972) and "Why Settle for Anything Less than Good Old-Fashioned Aristotelean Essentialism?" *Nous* (1973).

[23]*Tractus Logico-Philosophocus* (Routledge and Kegan Paul, 1961), 6.4311.

[24]To make this assumption plausible, we have to keep in mind disjunctive and degree-of-resemblance properties. Even with that complication, there may be further difficulties, but we shall ignore them for now.

Abortion

Listening to the Middle

Edward A. Langerak

Says one critic of the philosophical debate on abortion: "Philosophers are not listened to because they do not listen."[1] Though I believe the charge is too strong, my own review of the literature makes it uncomfortably understandable. If there is any public consensus on abortion, as reflected in legal systems as well as in public opinion surveys, it is the middle-of-the-road view that some abortions are not permissible but that others are, and that some of the permissible abortions are more difficult to justify than others. But many of the most widely cited philosophical writings on abortion argue that the only coherent positions tend toward the extremes: all or most abortions are put into the same moral boat with either murder or, more frequently, elective surgery. In fact, proponents of the extremes tend to respect one another as at least being self-consistent, while joining in swift rebuttal of those who want it both ways and ignominiously try to be moderates on either murder or mandatory motherhood.

This reaction against the middle derives from some basic beliefs of those on the extremes. On the liberal side are those who believe that fetuses, and perhaps even very young infants, lack some necessary condition (say, self-consciousness) of personhood.[2] This view is often combined with the further assertion that the social consequences of society's conferring on the fetus a claim to life are such that the conferral should not be made until birth or shortly thereafter. On the conservative side there are those who believe that from conception (or very shortly thereafter) the fetus has as strong (or almost as strong) a claim to life as does any person. This claim resides either in some property thought sufficient for personhood (say, genetic endowment) that the fetus has in itself, or in the immediate conferral of personhood on the fetus by God or society.

Of course, as Schopenhauer said, arguments are not like taxicabs that you can dismiss when they become inconvenient; and the two extremes are quick to point out the problematic implications of each other's positions. The liberals are accused of courting infanticide and the conservatives of trivializing the moral category of murder. Such implications would be more damaging to the extremes were it not that most moderate positions have an equally problematic flaw —that of arbitrary line-drawing. My reading of the abortion literature suggests that there are two widely shared beliefs that moderate positions seek to incorporate in their approaches to the abortion issue. The first belief is that something about the fetus itself, not merely the social consequences of abortion, makes abortions (or at least many abortions) morally problematic. The second is that late abortions are significantly more morally problematic than early ones. Not only are these beliefs widely shared by moderates, but I find that liberals and conservatives, whose positions implicitly reject one or both beliefs, often feel uncomfortable in rejecting them.

In accounting for these two beliefs, most middle positions maintain variations of what I call the "stage" approach and what its critics call the "magic moment" approach. The assertion is that at some point in the development of the fetus, say at the point of acquiring some vital sign, of sentience, of quickening, or of viability, the fetus suddenly moves from having no claim to life to having as strong (or almost as strong) a claim as an adult human. While the "stage" approach is consistent with the two beliefs underlying the moderate position, its difficulty has always been to explain the tremendous moral weight put on some specific point in what really amounts to a continuum in development. Critics on both extremes argue that, no matter what stage is picked as the "magic moment," the whole approach is *prima facie* arbitrary.

The implications of the liberal and conservative positions, including their denial of one or both of the moderate beliefs, and the *prima facie* arbitrariness of the stage positions, motivate consideration of an alternative that both is coherent and listens to the middle by accounting for the two beliefs.

Without examining all the alternatives, I will argue that the potentiality principle is plausible and accounts for the first belief—that something about the fetus itself makes abortion morally problematic—but that, by itself, it cannot account for the second belief—that late abortions are significantly more problematic than early abortions. I will then argue that a conferred claims approach is plausible, consistent with the potentiality principle, and accounts for the second belief though it cannot account for the first.

I will suggest that combining the potentiality principle with a conferred claims approach provides moderates with a coherent framework for thinking through the central questions of the abortion debate: (1) When does an individual human being attain either an inherent claim to life or such properties that society ought to confer on it a claim to life? (2) When do a person's or a group of person's claims to life, physical or mental health, freedom, privacy, and self-actualization override another human being's claim to life? (3) When should answers to the first two questions be incorporated into the law of a pluralistic society?

The Potentiality Principle

I formulate the potentiality principle as follows: "If, in the normal course of its development, a being will acquire a person's claim to life, then by virtue of that fact it already has some claim to life." To understand this principle, one must distinguish among "actual person," "a capacity for personhood," "potential person," and "possible person." An *actual person* is a being that meets a sufficient condition (whatever that may be)[3] for personhood and thereby has as strong a claim to life as normal adult human beings. Roughly, a *capacity for personhood* is possessed by any being not currently exhibiting that capacity, but who has proceeded in the course of its development to the point where it could currently exhibit it (for example, a temporarily unconscious person). A *potential person* is a being, not yet a person, that will become an actual person in the normal[4] course of its development (for example, a human fetus). A *possible person* is a being that could, under certain causally possible conditions, become an actual person (for example, a human sperm or egg).[5]

This technical set of distinctions is important because the potentiality principle asserts that potential persons, but not possible persons, have a claim to life. Some attacks on the principle confuse these categories.[6] Also, the principle is consistent with granting full personhood to those with a capacity for personhood, a fact ignored by those who collapse "capacity" and "potentiality" and argue, for example, that the category of "potential person" endangers sleeping persons. Moreover, the distinctions can help us avoid sloppy language, such as that of the Supreme Court in *Roe* v. *Wade* when it asserted that at viability the state begins to have a compelling interest in "potential life." Clearly a fetus is actually alive and is even an actual human being, genetically defined; its unique status is that, given most criteria of personhood, it is neither an actual person nor a merely possible person—it is a potential person.

Potentiality and Temporality

The potentiality principle asserts that a potential person has a claim to life, albeit one that may be weaker than the claim of an actual person. Many people find this assertion intuitively plausible, but are unable to persuade those who challenge it. Here is my attempt to persuade.

It is clear that the unique status of the potential person has to do with its inherent "thrust" or predetermined tendency. A potential person is not simply a set of blueprints, it is an organism that itself will become the actual person toward which it is already developing. Controversial issues of personal identity arise here, but two points seem obvious. First, we cannot simply assume that its predetermined tendency already grants it the claims it will have in the future. To paraphrase H. Tristram Engelhardt Jr., we must not lose the ability to distinguish between the claims of the future and those of the present[7]; or as S. I. Benn succinctly puts it, a potential president is not already commander-in-chief.[8] Second, those attracted to the potentiality principle do see some derivative relationship between the claims that a being will have in the normal course of its development and those that it has in the present.

I believe that the plausibility of the last point rests in perceiving humans as basically temporal beings. For actual persons this is true, first of all, from an internal point of view (a fact Heidegger uses for his entire ontology). Our self-consciousness so orients us to our past and our future that, in an important sense, we are our history and our projections as well as our present. A premedical student, for example, sees himself or herself as a future physician, not just as a science student. This temporal perception is also true from an external point of view, a point of view that extends to humans that are not yet persons. When we see a very young child, we see something of the adult it will, in the normal course of its development, become, as well as something of the baby that it once was. In this temporal perception lies, I believe, the respect we feel is due former persons (for example, respectful treatment of corpses) and, for that matter, former presidents. The respect we give former persons and presidents is not as great as that which we give actual ones, but that does not undermine the fact that some respect is due the former and that it is derivative from, indeed proportional to, the respect due to the latter.

Similarly, perceiving humans in a temporal context accounts for the respect many feel is due to humans by virtue of their potential. As an analogy, consider a potential president. Following my distinctions, such a person is not merely a possible president (something civics teachers used to say about every American child); he or she has already won the election but has not yet been inaugurated (on a somewhat arbitrarily selected date). The person is not yet commander-in-chief but, in the normal course, will (not could) be. Already that person receives some of the perquisites of the future office. The fact that the news media and others give the potential president more attention than the actual president, of course, may be the result of prudence, if not exploitation (the same derivation for much of the respect given actual presidents). But, at least in pre-Watergate times, some of the respect given actual presidents, and most of that given former presidents, derives from the high office that the person has or had, even when the person is not particularly deserving. Those who perceive a person in a temporal context and who, like myself, still respect an actual or former president by virtue of the office (apart from achievements in it), will derivatively have some respect toward a potential president by virtue of the office he or she will have.

Even those who deny that presidents ought to be respected simply by virtue of their office, should agree that some of the respect given persons derives from their "office" of personhood, apart from their achievements. In fact, traditionally the respect involving a claim to life derives from what persons are, rather than what they achieve or fail to achieve. If so, then perceiving humans in a temporal context should elicit some respect for former and potential persons, respect that is derivative from and proportional, though not identical, to the respect elicited by the actual persons they were or will become.

Temporality and Probability

Some may grant the strength of this argument as it applies to former persons, sensing that it accounts, for example, for our aversion to artificially keeping former persons "alive" in order to harvest their organs at a convenient time. However, whatever else we may say about former persons, they were certainly, at one time, actual persons. But the personhood of potential persons is still "outstanding" and there is no guarantee that it will be realized. The contigency of the "not yet" makes it asymmetrical with the "has been" even when we perceive humans in a temporal context.

This objection forces us to ask just what is the moral significance of the predetermined tendency of a potential person. Though the tendency does not guarantee personhood, it does distinguish the organism from possible persons by guaranteeing a dramatic shift in probabilities. This difference in probabilities is similar to that which distinguishes a potential president from a possible president. The potentiality principle asks us to respect a potential person by virtue not of what it *could* be, but of what it *will* be in the normal course of its development. Even those of us who refuse to mythologize the predetermined tendency in potential persons must agree that this tendency makes it highly likely that, without outside interference, they will become persons. Is this shift in probabilities of moral significance?

Consider the other end of the life-span. Those who believe that it is sometimes permissible to cease striving officiously to keep humans in an irreversible coma artificially alive, must agree that the irreversibility of the coma is seldom, if ever, absolutely guaranteed. But we believe it is morally irresponsible to allow the rare "miraculous recovery" to prevent acting on the best medical prognosis, when it indicates no reasonable hope of recovery. To shut off a respirator when there is a 50 percent chance of recovery, (or even a 5 percent chance, given our laudable bias toward erring in favor of personal life), is morally wrong, but not when the probability of recovery approaches (without reaching) zero. In an uncertain world, judgments of high probabilities are often the only kind we have. This makes dramatic shifts in probabilities morally significant.[9]

So I believe that the high probability of future personhood, inherent in a potential person, is of moral significance to those who perceive humans in a temporal context, and that this makes plausible the assertion of the potentiality principle. I hope I have at least shifted the burden of proof on to those who deny that the high probability of a fetus's becoming a person with a strong claim to life already grants it some (proportional) claim to life and respect.

Conferred Claims

Although the potentiality principle, as defended, accounts for the first belief—that something about the fetus itself makes abortion morally problematic—it leaves open the question of just how strong a claim to life should be attributed to the fetus. There are extreme liberals on the abortion issue who may grant the fetus some claim to life but simply argue that the claims of an actual person—claims to freedom and mental health—always override the claim to life of a fetus. Among those who use the potentiality principle, there will be intramural debates on how strong a claim to life it implies. I cannot argue the case here, but I believe that the most plausible use of it is one that allows the use of IUDs and "morning after" pills (both of which probably act as abortifacients), as well as abortions during the first trimester for such reasons as the woman's being too young for motherhood.[10] But then the claim to life attributed to the very early fetus cannot be very strong. The incidence of early spontaneous abortion is estimated[11] variously from 15 percent to over 50 percent, and second-trimester fetuses have a somewhat higher natural death rate than postviable fetuses. In other words, the probability of an older fetus becoming an actual person is perhaps double the probability of a zygote becoming a person. While this shift in probability is noteworthy, and marks implantation as a point of some moral significance, it is not nearly as significant as the difference in moral seriousness moderates see between a very early abortion and a late one. Consequently, if the inherent claim to life of a potential person is derived from and proportional to the probability of its becoming an actual person, one cannot in good faith allow the claim to life of a zygote to be easily overridden and then assert that the inherent claim to life of an older fetus is so vastly stronger that it all but cannot be overridden. Therefore, although the potentiality principle can account for the belief that something about the fetus itself makes abortion morally problematic, it cannot by itself account for the belief that late abortions are significantly more morally problematic than very early abortions.

However, the conferred claims approach can account for the second belief, although it cannot account for the first. Assume that, whatever moral claim to life an older fetus may have by virtue of its potentiality, the claim may not be strong enough to override the claim of a pregnant woman for an abortion. At what point should society confer a stronger claim to life on the fetus? At what point should society treat it as if it were a person?

The conferral approach to the status of the fetus is not an unusual one,[12] though it is sometimes thought incompatible with an approach that asserts an inherent claim in the fetus itself. But an approach that *confers* claims rubs an approach that *recognizes* inherent claims only if the inherent claim to life is thought to be as serious as an actual person's claim to life. In this case it would be futile (rather than contradictory) to ask what claims society ought to confer on it. However, when the recognized inherent claim is weaker than a normal adult's claim to life, as can be the case with the potentiality principle, one can coherently ask whether society ought, in addition, to confer on the fetus a stronger claim to life.

The argument in favor of such a conferral basically appeals to the social consequences of abortions and infanticide. For example, infants are so similar to persons that allowing them to be killed would generate a moral climate that would endanger the claim to life of even young persons. And older fetuses are so similar to infants that allowing them to be killed without due moral or legal process would endanger infants. Of course there must be a cutoff for this sort of argument. For example, most would agree that preventing the implantation of zygotes would have no discernible effect on our sympathetic capacities toward persons. At what point would abortions begin to have such effects, especially on medical personnel, that it is in society's interest to endow the fetus at that point with a stronger claim to life? This seems largely an empirical question and one not easily answered,[13] though I will suggest some guidelines below.

One difficulty with the conferral approach has always been that the relevant considerations are the interest and sympathies of actual persons, rather than moral claims inherent in the fetus itself. Indeed, the above argument is reminiscent of Kant's view that we ought not beat our dogs merely because beating our dogs might make us more inclined to beat people. Such arguments derive protection for some beings from the rather variable, even capricious, sympathies of other beings. Thus the conferral approach by itself does not account for the belief that something about the fetus itself makes abortion morally problematic; but this belief is accounted for by the potentiality approach.

Implantation, Quickening, Viability, and Birth

My combined approach escapes the problematic implications of the extremes but does it escape the flaw of arbitrary line-drawing that I attributed to those moderate positions that appeal to the stage or "magic moment" approach? Two related considerations show that it does. First, notice that the word "arbitrary" should not be used loosely. For example, there is a certain arbitrariness in making 18 the age of majority rather than 17 or 19. But the relevant criteria nonarbitrarily imply that, if a legally precise line must be drawn within the continuum of growth, the debate must focus on that time span rather than, say, the span between seven and nine.

Second, I submit that the two criteria I use—important shifts in probabilities and dangerous social consequences—nonarbitrarily suggest four spans (beyond that of conception) for moral and legal line–drawing in a potential person's continuum of growth. Although these criteria imply distinct spans for definite increments in the strength of the claim to life, at no stage does a potential person move from having no claim to having one as strong as an adult.

The first span, as we saw, is that of implantation, when the shift in probabilities of actual personhood signifies a somewhat stronger inherent claim to life, at least from the moral point of view. The recognition of this change is due apart from any consequentialist considerations about the difference between more or less unknowingly preventing implantation and knowingly detaching an implanted embryo. However, the remaining spans are suggested by consequentialist considerations about the psychological and social impact of abortions, considerations in favor of conferring an even stronger claim to life on the fetus.

The second span involves the traditional indicator of "quickening." When the fetus begins making perceptible spontaneous movements (around the beginning of the second trimester), its shape, its behavior, and even its beginning relationship with the mother and the rest of society (every father recalls when he first felt the fetus's movements) all suggest that abortions after this point will have personal and social consequences specifiably more serious than those of earlier abortions.

The third is that of viability, when a fetus is capable of living, with simple medical care, outside the womb (around the end of the second trimester). Recall the "infanticide" trials of physicians who, claiming they were inducing abortions, were charged with participating in premature births and murders. This controversy is only one indication that killing potential persons after viability has social consequences (apart from legal ones) even more serious than abortions soon after quickening.

Finally, consider that allowing infanticide is generally regarded as a *reductio* of those positions that allow it. The aversion to infanticide is shared even by most of those whose criteria for personhood imply that a newborn is still only a potential person and not an actual one. This suggests that most people agree that at birth the potential person attains properties and relationships so close to those of actual persons that the consequences of killing at this point are practically the same as killing young persons.

If these observations are true, they justify conferring on newborns a claim to life as strong as that of adult persons. They also suggest partial wisdom in the Supreme Court's decision allowing states to grant a rather strong claim to life to postviable fetuses, a claim overridden only by the claim to life or health (I would specify "physical health") of the mother. But the court decision, in effect, mandates the allowing of abortion on demand for all previable fetuses. If my observations about quickening are correct, we should also draw an earlier line, conferring a claim to life on the fetus at the beginning of the second trimester, a claim less strong than that conferred at viability, but one overridden only by such serious claims as that of the mother to mental or physical health.[14] Probably the moral line drawn at implantation should remain outside the legal realm.

I admit the difficulties in legally implementing such an approach, but I doubt that they are insurmountable or as deep as the moral and legal difficulties of alternative approaches. Therefore I believe I have presented a plausible approach to the abortion issue that is coherent, is not arbitrary, and listens well to the considered intuitions of those in the middle.[15]

References

[1]Roger Wertheimer, "Philosophy on Humanity," *Abortion: Pro and Con,* ed. Robert L. Perkins (Cambridge: Schenkman Publishing Company), p. 127.

[2]For brevity I use "fetus" in a generic sense to refer to unborn humans at any stage of development, including that of zygote, conceptus, and embryo. I assume the fetuses are human beings, genetically defined, and use "person" to refer to those human beings that have as strong a claim to life as a normal adult. I use "as strong a claim" rather than "same claim" because, if very young human beings are persons, their claim to life clearly involves the claim to be nurtured as well as the claim not to be killed, a feature that is not clearly true of a normal adult's claim to life. I use "claim to life rather than"right" or "prima facie right" because my argument entails that a fetus's (though not a person's) claim to life can be held with varying degrees of strength, and I agree with Joel Feinberg (*Social Philosophy,* Englewood Cliffs: Prentice Hall, 1973, pp. 64–67) that this is a feature of claims rather than rights. Though Feinberg may object to my use of his distinction, I agree with him that the "right" or "valid claim" in a given instance is the strongest of competing claims. For an account of the relationship between claims and rights that I believe is consistent with my argument, *see* Bertram Bandman's "Rights and Claims" in *Bioethics and Human Rights,* eds. Elsie L. Bandman and Bertram Bandman (Boston: Little Brown and Company, 1978).

[3]One advantage of the potentiality principle is that one need not specify the necessary or sufficient conditions for actual personhood; one need only note that, whatever they are, a potential person will acquire them in the normal course of its development. My own position is that self-consciousness is a necessary and perhaps a sufficient condition for personhood: "The fact that man can have the idea 'I' raises him infinitely above all the other beings living on earth. By this he is a *person*" (Immanuel Kant, *Anthropology from a Pragmatic Point of View,* trans.

Mary J. Gregor, The Hague: Martinus Nijhoff, 1947, p. 9). *See* also H. Tristram Engelhart Jr. ("The Ontology of Abortion," *Ethics* **84**/3, April, 1974, 230 n): "Only self-conscious subjects can value themselves, and, thus, be ends in themselves, and, consequently, themselves make claims against us." While Joel Feinberg seems to object to thinking of personhood as a property, he does appeal to the fact that persons are "equally centers of experience, foci of subjectivity" (*op. cit.*, p. 93).

[4]Although using the phrase "in the normal course of its development" rather than "in the normal course of events" emphasizes the teleological ("nature's aim") rather than the statistical probability aspect of "normal development," my later argument about probability and claims assumes that even a teleological notion of "normal" has statistical implications: if the natural end of (a) is to become (A) then it is highly probable that, without interference, (a) will become (A): I believe I am referring to what some Thomists call "active, natural potentiality," though I deny potential personhood is as claim–laden as actual personhood.

[5]The class of potential and possible persons must be distinguished from the class (membership unknown) of future persons, namely the class of future actual persons who do not now exist but will in fact exist in the future. One must be careful with analogies between our duties to potential persons and our duties to future persons (for such an analogy, *see* Werner S. Pluhar, "Abortion and Simple Consciousness," *Journal of Philosophy* **74**/3, March 1977, 167). If there are future persons (as is so likely as to be certain), they will be actual persons whose quality of life will be affected by actions we now perform, while it is debatable whether killing potential persons affects the quality of their lives *as persons*.

[6]A point I argue in reply to Michael Tooley's "Abortion and Infanticide," *Philosophy and Public Affairs* **2**/4 (Summer, 1973), 410–416.

[7]Engelhardt, p. 223.

[8]"Abortion, Infanticide, and Respect for Persons," *The Problem of Abortion*, ed. Joel Feinberg (Belmont: Wadsworth Publishing Co., 1973), p. 103.

[9]In this I agree with John T. Noonan, "An Almost Absolute Value in History," *The Morality of Abortion*, ed. John T. Noonan, Jr. (Cambridge: Harvard University Press, 1970), though he seems to argue wrongly that an abortion involves a high probability of killing a person. Instead, it kills a human that had a high probability of becoming a person.

[10]Notice that if one uses the potentiality principle to attribute a very strong claim to life for the fetus, one has, in effect, denied the belief that late abortions are significantly more morally problematic than early abortions.

[11]*See* Malcolm Potts, Peter Diggory, and John Peel, *Abortion* (Cambridge: Cambridge University Press, 1977), chap. 2. The highest estimate I have seen, at variance with most others, is 69 percent, by Harvard physiologist John D. Biggers (*Science* **202**, October 13, 1978, 198).

[12]*See* R. B. Brandt, "The Morality of Abortion" (*The Monist* **56**, 1972, 504–526), for a quasi-Rawlsian development of this approach. *See* also Ronald M. Green, "Conferred Rights and the Fetus," *Journal of Religious Ethics* **2**/1 (1974), and Benn (*op. cit.*)

[13]*See* Magda Denes, *In Necessity and Sorrow: Life and Death in an Abortion Hospital* (New York: Penguin Books, 1977), for one description of the different social effects of abortions at different stages of pregnancy.

[14]Notice that adding the conferred claims approach highly qualifies a possible implication of my defense of the potentiality principle, namely the implication that it is somewhat easier to justify aborting fetuses with defects that lower their probability of attaining personhood. Any arguments for conferring a stronger claim to life on fetuses at a given point would apply to most defective fetuses as well.

[15]Patricia Fauser, James Gustafson, Gary Iseminger, Daniel Lee, and Frederick Stoutland gave me very helpful comments on an earlier draft of this essay.

The Life of a Person

Roland Puccetti

When one reflects on the life of a person, it becomes immediately apparent that this can be done in two very different ways. One way is to look upon the person as a particular organism with a spatio-temporal history of its own; there the identity question is approached from the outside, so to speak, and differs not at all from questions about the identity of material objects through time. The other way is to look upon the person's life history as the total span of conscious experience that this person has had; here identity is approached from the inside, whether it be your own life you are reflecting on, or that of another person. It was Lucretius' contention, in Book III of *De Rerum Natura*, that since good or harm can accrue only to a subject of conscious experiences, the latter is the correct view and the former leads to superstition.

I am inclined to think Lucretius was right about this, that personal life is ineluctably shorter at both ends than the life of the organism which is the biological substrate of the person, because developmental processes at the beginning and degenerative processes at the end of organic life are insufficient to support conscious functions, and without these nothing done to the living tissue has personal value or disvalue. But to bring out the intuitive force of Lucretius' position, consider the following parable.

265

1. The Genie's Bargain

Suppose that one day a genie appears before you and convincingly displays magical powers. Now he tells you that he will, if you want, expand your brain so that you will have an IQ of 400. This means that whatever you're interested in achieving will come within easy grasp, whether it be leadership of state, heading a conglomerate, outpainting Picasso, or getting a Nobel Prize in medicine. However, he explains, there is one hitch. If you want him to do this for you, you will at the moment of brain expansion cease forever to have conscious experience. No one will know this, for he will program your brain to make all the correct responses to questions, etc. Nevertheless the great future achiever will be an automaton and no more. Would you accept this offer?

I have tried this out on several groups of students and it is amazing how uniform the reaction has been. A few say they would accept the bargain, but upon probing it turns out they have in mind an altruistic act that might lead to discovering a cure for cancer or building world government. They, as well as the majority who refused, all agreed that accepting would be tantamount to personal annihilation, at least in this world, and that one could indeed doubt that wholly unconscious achieving automata with their bodies would still be *them*. But if this is correct, and given that the one and same living organism persists from conscious to permanently unconscious activity, it appears bodily continuity through time is not a sufficient condition of personal identity, whereas continuity of consciousness is, and that without even a capacity for conscious experience there is no person any more.

Let us now apply the same lesson to the other end of the life spectrum. Suppose the genie tells you that you are the reincarnation of Napoleon Bonaparte, whose life you happen to admire greatly. You protest that this can't be, since you have no recollection of doing the things Napoleon did, or experiencing his triumphs and defeats. The genie replies that of course you do not because as a young man Napoleon accepted the genie's bargain and thus did all he is remembered for in the history books quite unconsciously. Just supposing you believed this, would you feel proud of Napoleon's deeds as if they had been yours? I think not, because without any sense of those deeds being included in your personal conscious history, they could just as well be the past deeds of anyone else around. Once again it is psychological continuity which underlies the strand of personal identity, and in its absence there is no clear notion of being one and the same person.

But if so, how vain it seems to extend personhood beyond the loss of capacity for conscious experience, and equally so to thrust it back in time to a stage of organic life before that capacity existed. Yet, as we shall now see, this is exactly what many people tend to do.

2. Possible Persons

I begin with a notion Lucretius certainly would have found as strange, namely that of *possible* persons. Here the reference is not to something so general as, say, persons who will live five generations from now, but to *specific* as yet unconceived humans. R. M. Hare [6], for example, asks whether a life-saving operation for an abnormal child might not be denied so that the parents, facing the burden of caring for such an offspring, will not be discouraged from having normal, healthy children later on. He invites us to imagine "the next child in the queue", whom he christens "Andrew," and asks if a full, constructive, and probably happy life for this possible person is not a better moral outcome than sustaining the abnormal child and risking Andrew's future nonexistence. Hare grants that since Andrew is only a possible person he cannot be *deprived* of life, but suggests that he can be harmed by *withholding* life from him. And Derek Parfit [11], commenting on Hare's remarks, apparently concurs. He asks why, if it can be in a person's interest to have his life prolonged, it cannot be in his interest to have it started.

To this the Lucretian response would surely be that life cannot be *withheld* from a nonexistent subject any more than this nonexistent subject can be deprived of it. Similarly, a nonexistent subject cannot have his life *started* by anyone. To talk this way is to imagine there are ghostly persons somewhere just waiting to be given flesh and blood, but there are none. For suppose one agreed with Hare that it is unjust to withhold life from Andrew. In that case, how could one make restitution to him? Where would we go to find this possible person and give him, at last, the life he deserves? All one can do is imagine that the parents of the abnormal child, once it is gone, might conceive a healthy child two years down the road and, when it is born, baptize him "Andrew." It is only by retroactively predating this latter child's existence beyond the point of conception that we get the notion of him, quite illicitly, as a specific possible person awaiting conception. There are no such persons (except in the barren sense of a *logically* possible combination of genes occuring in the conceptus), and thus no harm can be done them.

3. Potential Persons

When I was writing on this topic many years ago [14], I used the term *potential* persons to refer to human children, for reasons I shall make clear later. However, I now find the term has been preempted in the literature to refer to fetuses, and shall follow that usage here. Fetuses have an advantage over possible persons in that they exist; the question is whether they are really the sorts of entities that qualify as having a right to life (assuming there are *any* such entities) by virtue of their potential for becoming human persons.

Michael Tooley [15] has argued strongly against the potentiality principle as follows. Most of us would agree that it is not morally wrong, or only slightly so, to destroy surplus newborn kittens. Now if a chemical were discovered that, when administered to newborn kittens, led to their developing into rational, language-using animals and hence candidate persons, would it become grossly immoral to continue destroying newborn kittens? If not, and if artificial vs natural potentiality for becoming persons is not a morally relevant distinction, then naturally potential persons have no more right to life than such kittens would have.

What is it like to *be* a potential person, i.e., from the inside of the stage of life? None of us knows, not because we have forgotten what it is like, but because our conscious personal lives had not begun yet. I do not deny that it is possible a fetus has crude sensations of pressure, temperature, etc. But zygotes, morulae, blastocysts, embryos, and even (in the technical sense) early 'fetuses' do not have the neural complexes necessary to sustain believably a conscious and therefore, by Lucretius' criterion, a personal life. If not, no *person* begins his or her personal life before late term in the intrauterine environment, and only barely then. Terminating a human life before that stage is not, therefore, killing an innocent *person*. It is destruction of at most the organic blueprint of a future person.

For suppose we had the means, technically, of saving life and promoting normal development of a spontaneously aborted fetus at *any* stage whatever. Would the zygote have *more* life ahead than a near neonate? Biologically, yes. Does that mean more personal value accrues to the saved zygote than to the premature baby? Surely not, for if they both lived a normal human life span and had equal enjoyment of it, it matters not at all that the zygote's organic life was saved seven or so months earlier; at that time its personal life had not yet begun and nothing in the events of those months adds to that future person's enjoyment of life. If so, can any event detract from it?

Many would say yes, for abortion constitutes an abrupt cancellation of the promise of a future personal life. My point is that cancellation of a promise is not cancellation of the thing promised. Take a young couple who have two healthy, happy children. It may be that at the time of the second pregnancy they gave serious thought to terminating it as inopportune, but relented and now are glad they did, for he is a wonderful child who shows every sign of enjoying a long and properous life. The temptation is to say they are glad not only to have him as a son but glad they did not deprive *him* of his life. But on the Lucretian stance I am developing here, the latter source of self-contentment is confused. He did not exist as a *person* until shortly before his birth. Abortion of the fetus from which his personal life ensued would have prevented someone with his particular genetic throw of the dice from getting launched as a person, but could not have ended a personal life, for this had not yet begun. Blueprints and miniature models are not edifices. What is more, all those early formative experiences in his personal life, comparable to the architect's dabbling with the original design to secure improvements as it actually takes shape, are indispensable ingredients in the individuation of the growing structure of a conscious human, and none of those could have taken place before extra-uterine life, so the *particular person* he is would not have lost this actual life.

4. Beginning Persons

I will speak now of *beginning* persons, as a term to replace the preempted 'potential persons' I used to designate human neonates and infants more than a decade ago. The reason I did this was that I then wanted to reserve 'person' as coextensive with *moral agent*. A moral agent, I said, is both moral subject and moral object, and while small human children are moral objects, as are other higher forms of animal life, by virtue of being able to suffer, they are not yet moral subjects but only potentially so. Adapting now to the terminological shift, I would say that potential persons, meaning fetuses before late term, are not even moral objects,[1] whereas beginning persons are, just as they are potential moral subjects as well.

But Tooley [15] has questioned even this. He holds that it is only when an organism becomes *self*-conscious and has a concept of itself as a continuing subject of experiences that it qualifies as a person with a right to life. Such a view, if correct, could be used to justify infanticide as well as feticide, on grounds that without linguistic abilities normally not developed before the second or third year of life, human infants do not have a self-concept. However, Tooley recognizes that it is possible a nonverbal concept of self emerges as early as a month after birth, pushing back the 'cut-off point' between potential and actual personhood to the first few postnatal weeks; but then he worries that if there *is* such a nonverbal self-concept all kinds of infra-human species devoid of language functions might also qualify as persons whose right to life we humans routinely override.

Let us see what can be salvaged here. If it were true that without a verbal conceptual scheme no self-concept is possible, would this imply that beginning persons have no right to life (assuming, again, that any entity has this)? Consider, first, that there are some otherwise normal children who were raised in isolation by uncaring parents, cut off from human language, who if not rescued by the age of 10 or 12 are thereafter unable to learn language, even in an artificially enriched linguistic environment. Conversely, consider that some higher primate species who never develop symbolic language in natural conditions have been trained in similarly enriched environments to do so using plastic cutouts, computer consoles, and American Sign Language. If what Tooley suggested were true, it follows that the isolated, otherwise normal human 12-year-old unable to learn to talk has no right to life, whereas the chimp who can sign "You take out cabbage and give me monkey chow" is a person with a right to life! Yet what morally relevant difference is there between the isolated child and the run-of-the-mill infant toddler? The fact that the former is artificially speechless and the latter naturally so cannot serve to distinguish between them. And if not, the lack of a verbal self-concept in beginning persons cannot justify denying them whatever right to life anyone has.

5. Actual Persons

Tooley's qualms about the person-status of languageless infrahuman species can now be addressed. The only solid evidence for a nonverbal self-concept comes from Gordon Gallup's [4] studies of self-recognition in a reflecting surface by higher apes, something lower apes such as monkeys cannot learn to do. Yet monkeys rely on recognition of *other* monkeys' faces to establish their place in the troop's dominance hierarchy. How can this be? After all, brain-damaged humans who lose the ability to recognize faces not only cannot identify their own but cannot recognize those of close friends or loved ones. Apparently it is because chimpanzees and other higher apes have a self-concept to begin with that they quickly learn to recognize their own faces and bodies in a mirror. Lacking this, lower apes such as the monkey can identify and react to other monkey faces appropriately, but persist in seeing the reflected face and body as that of just another conspecific of similar age, sex and size. So if it were true that regarding oneself as a continuing subject of experiences is what qualifies an organism for personhood and the right to life, on present evidence only our closest phylogenetic relatives would make the grade.

But even this seems strained. According to many contemporary philosophers, the kind of rude nonvocal but still verbal abilities demonstrated by chimpanzees after arduous human training, plus the evidence for a nonverbal self-concept already in place in such species, is a far cry from what *actual* persons like you and me are able to do. For example, H. Frankfurt [3] has influentially espoused the view that a necessary condition of being a person is the capacity for having what he calls 'second-order volitions,' i.e, the desire not to will what one wills and will something else instead. And this clearly requires a rich verbal conceptual scheme, for how else is one going to think the equivalent of, "I wish I were less ambitious," or "If only

I could love her in return?" Beings that do not have this ability are simply characterized by Frankfurt as 'wantons,' and include nonhuman animal species, mental defectives, and the small children I called beginning persons.

More than a decade ago, I would have welcomed Frankfurt's stipulation, because at that time I could not see how any entity could be a moral subject as well as a moral object, hence a moral agent and a full-blown person, without a complex verbal conceptual scheme. But since then I have come to distrust such maneuvers by philosophers, for the reason that they are dangerously exclusive. Who am I to say, for example, that someone with a lesion to Broca's area, a motor aphasic unable to think in propositional language anymore, is therefore a mere 'wanton,' a nonperson without a right to life? Except in the narrow legal sense that such a human may not sign a contract or witness a will because of linguistic incompetence subsequent to the brain damage, I might indeed prefer to regard such a human as a person with a language defect, no more and no less than that. And in that case I would have to conclude that moral agents are just a subclass of persons.

Then what is a person? I have come to share the skepticism of D. C. Dennett [1] over ever being able to give an exhaustive list of the necessary and sufficient conditions of being a person; as he says, it might turn out that the concept of a person is only a free-floating honorific that we all happily apply to ourselves and to others as the spirit moves us, rather as those who are *chic* are all and only those who can get themselves considered *chic* by others who consider themselves *chic*. In any case, it is not my task here to say exactly what a person is, but only to argue that a person is more than a living human organism[2]; it is a conscious entity that builds a personal life from agency and experience, and until and only for so long as it has a capacity for conscious experience does the notion of a right to life, if there is such a thing, take hold.

6. Former Persons

Probably there has been no time before this when philosophers have been more conscious of the brain dependence of the human mind. Yet in spite of this, they sometimes talk of the brain as if it were a replaceable or substitutable organ a person has, on a par with the heart, a kidney, or the cornea of the eyes. For example, John Perry [13] has suggested that some day it might be possible to make a duplicate 'rejuventated' brain exactly like the original except for having healthy arteries, etc., which could then be used to replace the latter when it starts to wear out. If the copy were exact enough, he argues, all the individuating psychological characteristics of the person, including the long-term memories he has, would persist and exact similarity is as good as makes no matter to saying one and the same person has survived the operation. John Hick [7] has even extended this notion to the next world. He asks us to imagine that upon our earthly demise God will create a replica of each of us in a special Resurrection World, complete with an exactly similar brain containing the same memory traces, dispositions, etc., and holds that it would be unreasonable for these resurrectees not to regard themselves, and be regarded by others there, as continuants of the persons whose earthly pasts they recall as their own.

What such claims overlook is that for any future person to be me, any statement true of me now would have to be true of him as well, otherwise he is not me. Now it is true of me that I can really remember certain milestone events in my life, e.g., a delayed honeymoon on the island of Corfu. But the person with the duplicate of my brain came into existence as a subject of experiences only upon duplication, just as the replica of me in the Celestial City would come into existence upon my death, not before. But then neither the duplicatum nor the replicatum of me, given that the brain each has is what makes each a subject of experiences, could possibly have been a subject of experiences at the time of my honeymoon in Corfu, and so could not really remember the events there, but only seem to. If so, neither would be me and it would be vain for me to anticipate any experiences they are going to have as experiences I shall have. Duplication or replication of a brain cannot endow the resultant person with retroactive personal history; not even God can change the past and tomorrow make true of it what was not true of it before. In sum, it is the spatiotemporal continuity of a particular living brain that is the anchor of personal identity through time.

If we have this straight, we may now ask when exactly does a person's life end and he or she become only a *former* person? At the beginning of this paper, I suggested that on the Lucretian view personal life spans one's total conscious experience and that this is necessarily shorter at both ends than the life of the organism supporting that conscious life. We have seen how this is probably so for the first several months of fetal development, but one might well wonder if it is true at the end of organic life in any more than a picayune sense. After all, if the organic basis for conscious life is the brain and total brain infarct subsequent to, say, cardiac standstill or lung failure causes the death of masses of central neurons by oxygen deprivation within, normally, a matter of minutes, organic death of the brain follows very quickly upon loss of consciousness. It is true that electrical activity can persist in the spinal cord for hours, and some somatic cells may take up to two days to die, such as those composing cartilage in the knee, but these lingering signs of life are no obstacle to a medical finding that the person has died.

Such is indeed normally the sequence of events, but not always. Consider the following case reported by Ingvar et al. [8].

> Case 8. The patient (Th. Sv.) was a female who had been born in 1936. In July 1960, at the age of 24, she suffered severe eclampsia during pregnancy with serial epileptic attacks, followed by deep coma and transient respiratory and circulatory failure. In the acute phase, Babinski signs were present bilaterally and there was a transitory absence of pupillary, corneal and spinal reflexes. A left-sided cartotid angiogram showed a slow passage of contrast medium and signs of brain edema. An EEG taken during the acute stage did not reveal any electrical cerebral activity. The EEG remained isoelectric for the rest of the survival time (seventeen years). After the first three to four months the patient's state became stable with complete absence of all higher functions.
>
> Examination ten years after the initial anoxic episode showed the patient lying supine, motionless, and with closed eyes. Respiration was spontaneous, regular, and slow with a tracheal cannula. The pulse was regular. The systolic blood pressure was 75–100 mm Hg. Severe flexion contractures had developed in all extremities. Stimulation with acoustic signals, touch or pain gave rise to primitive arousal reactions including eye opening, rhythmic movements of the extremities, chewing and swallowing, and withdrawal reflexes. The corneal reflex was present on the left side. When testing was done on the right side, transient horizontal nystagmus movements were elicited. Pupillary reflexes were present and normal on both sides. On passive movements of the head, typical vestibulo-ocular reflexes were elicited. The spinal reflexes were symmetrical and hyperactive.

Patellar clonus was present bilaterally. Divergent strabismus was found when the eyes were opened [by the examiner]. Measurement of the regional cerebral blood flow on the left side (ten years after the initial anoxic episode) showed a very low mean hemisphere flow of 9 mL/100 g/min. The distribution of the flow was also abnormal, high values being found over the brain stem. The patient's condition remained essentially unchanged for seven more years and she died seventeen years after the anoxic episode after repeated periods of pulmonary edema.

Autopsy showed a highly atrophic brain weighing only 315 grams. The hemispheres were especially atrophied and they were in general transformed into thin-walled yellow-brown bags. The brain stem and cerebellum were sclerotic and shrunken. On the basal aspect some smaller parts of preserved cortex could be seen, mainly in the region of the unci. Microscopically the cerebral cortex was almost totally destroyed with some remnants of a thin gliotic molecular layer and underneath a microcystic spongy tissue with microphages containing iron pigment. The white matter was completely demyelinated and rebuilt into gliotic scar tissue, and there were also scattered macrophages containing iron pigment. The basal ganglia were severely destroyed, whereas less advanced destruction was found in the subfrontal basal cortex, the subcallosal gyrus, the unci, the thalamus and hypothalamus, and in the subocular and entorhinal areas. In the cerebellum the Purkinje cells had almost completely disappeared and were replaced by glial cells. The granular layer was partly destroyed. The cerebellar white matter was partly demyelinated. In the brain stem some neurons had disappeared and diffuse gliosis was found. Several cranial nerve nuclei remained spared. The long sensory and motor tracts were completely demyelinated and gliotic, whereas transverse pontine tracts remained well myelinated ([8], pp. 196–198).

This clinical picture, confirmed by the autopsy findings, goes by various titles in the literature: cerebral as opposed to whole brain death; neocortical death without brain stem death, and more recently and appropriately, as 'the apallic syndrome,' because the characteristic feature is selective destruction of the paleum, the cortical mantle of gray matter covering the cerebrum or telencephalon. As it happens, the neurons composing the paleum are the most vulnerable to oxygen deprivation during transient cardiac arrest or, as in the above case, asphyxiation. Whereas with whole brain death, therefore including the brain stem that mediates cardiopulmonary functions, the patient can be maintained on a respirator only up to a week in adults and two weeks in children before cardiac standstill, the apallic patient can breathe spontaneously and demonstrate cephalic reflexes, which are also brain- stem mediated, for months and even years if fed intravenously and kept free of infection, thus allowing organic recovery after the top of the brain is gone.

I said 'organic recovery' is possible with destruction of the cerebral cortex; but on the Lucretian model *personal* life thereupon comes to an end, for with the paleum gone the very capacity for conscious experience goes as well. That such was indeed the case with this patient is obvious from the time of stabilization a few months after the anoxic episode: how else can one explain the persistently flat EEG, the inability to move even the eyes voluntarily, the reduction of cerebral blood flow to less than 20% of normal, and the spastic flexion of extremities? Thus 'the *patient*' was nonsentient and non-cognitive for seventeen years, but the *person* was not, for she had died all that long ago, and what was left was a still breathing *former* person.

How can one be sure? Perhaps a homely analogy will help the medically uninitiated to understand this. Suppose we wanted to find out if anyone lives in an apparently abandoned house, on the top floor. But we dare not break into it to see, for legal and ethical reasons. So we stand outside, watching and listening. We can hear the furnace go on, but that could be because of an automatic thermostat. We also see the lights go on in the evening, but that could be the result of an automatic timer to thwart burglars. We dial the phone number and hear the instrument ringing, so the lines are still intact, but no one answers. We measure the heat flow from the furnace and find not enough is reaching that top floor to keep any occupants alive there in winter. Finally, we attach listening devices to the outer walls and video cameras to the windows, but absolutely no real activity is picked up. Surely at this juncture, we would conclude it pointless to go on fueling the furnace and scrubbing the walls. Nobody is home upstairs.

Yet as things now stand, so long as the furnace goes on and the lights light up by themselves, we are supposed to be committed to heroic maintenance measures tying up scarce medical resources, even though there is *no one* being helped by these efforts. Lucretius would call this rank superstition and advise us to dispose of such former persons as reason dictates. After all, he would surely say, unconscious breathing and heart beating has no intrinsic value to a departed person; you could do no more harm to *that* individual, now dead, than you could do by opening a grave and stabbing a corpse. And I think he would be right.

No doubt this hard Lucretian line will appall many hearing it for the first time. I shall close by anticipating objections and trying to defuse them in advance.

7. Objections and Replies

Objection: Current legislation in the United States, and apparently Canada is following suit [10], is in accord with the 1968 Harvard Statement, which allows a determination of death subsequent to the *whole* brain dying above level C_1, as evidenced by prolonged absence of cephalic reflexes and of spontaneous heart and lung activity. What you are suggesting is presently proscribed by law.

Reply: Superstitious attitudes often get enshrined in law. Why not change the law? It's been done before.

Objection: It is unconscionable to prepare a patient for burial who is not apneic.

Reply: Spontaneous breathing is normally promissory of a return to conscious functions; with the apallic patient it is not. You can always stop the breathing.

Objection: That's euthanasia, whether passive or active, and it's illegal.

Reply: You're still confusing the patient with the person. If euthanasia means 'mercy killing,' it has no application here. How can you be *merciful* to someone already long beyond any possible suffering?

Objection: Nevertheless such actions would harden medical people. How do we know they won't just go through the wards disposing of helpless persons, such as mental defectives, the recoverably comatose, and the senile?

Reply: It is not because the apallic patient is helpless that I am recommending disposal, but because he is a dead person. The categories you mention retain a capacity for conscious experience, even if diminished, and to dispose of them would be to deprive them of the rest of their personal lives. You can still harm such people.

Objection: You take it for granted that the paleum is the seat of consciousness and a personal life. Yet Wilder Penfield [12], that great explorer of the cortex, believed to the end of his days that it was only a way-station and that the true site of personal being is central gray matter in the upper brain stem, which in the patient you referred to seems to have been well preserved. How can we be sure such patients are not secretly conscious?

Reply: On this issue Penfield was wrong. Split-brain surgery for relief of epilepsy yields independent streams of consciousness in the disconnected hemispheres, yet the brain stem is untouched.

Objection: Still, the autopsy showed a fairly well preserved hypothalamus and thalamus. Would this not indicate that the patient could feel thirst and pain?

Reply: *Who*, exactly, would be feeling pain or thirst? Someone who cannot remember, dream, think, anticipate, or come into contact with reality? But even supposing our concept of a person could be reduced to sensory islands like these, what harm would accrue to such an individual if he or she were prevented from experiencing further thirst and pain? In my own case, were this my "personal" future, I would prefer to go without it; and so too, I think, would any rational person.

Objection: The clinical picture included withdrawal reactions to noxious stimulation.

Reply: Which can be mediated by spinal arc pathways alone, as has been shown by experiments with paraplegics [5].

Objection: But if there's any uncertainty at all, why not give the patient the benefit of doubt? There have been misdiagnoses of even total brain death. It is always better to mistakenly assist a dead person than to mistakenly abandon someone who might otherwise survive.

Reply: The only misdiagnoses of total brain death I have heard of involved reduced oxygen requirements due to hypothermia or barbiturate intoxication. But of course one must be cautious. Take six months, use four-vessel angiography, the bolus technique, measurement of regional cerebral blood flow, EEG, brain scan, everything. When there is no longer room for doubt, there is no longer reason for concern.

Objection: Lucretius was an atheist. Why should a religious physician accept that personal annihilation is the result of the cessation of any particular bodily function?

Reply: You could say the same for whole brain death. What matters is not whether the soul survives the body and goes on to another world, etc., but the point at which a person's life ends in *this* world.

Objection: This is not so much an objection as a query. How is it that organic life can exceed conscious life so long, whereas a similar picture at the beginning of life is quickly fatal? I mean the anencephalic infant. Although the head is flattened, brain stem reflexes and spontaneous heart and lung activity are present, as in the apallic syndrome, yet the infant dies within weeks or at most a few months.

Reply: There are anecdotal references to one case of anencephaly where the child, under the total care of the mother, survived beyond seventeen years [9]. Nature is cruel, and a fairly long organic human life can still preclude even beginning a personal life.

Notes

[1]However, Engelhardt [2], whose analysis is fully supported by my own, believes that there is sufficient evidence to indicate the aborted fetus feels pain ([2], p. 334). If so, it would be a moral object and I am wrong to think otherwise. Yet being a moral object is obviously not a sufficient condition of being a person, as Tooley's example [15] of surplus newborn kittens makes clear. The reason I hesitate to ascribe a 'right to life' unreservedly even to human beings is that I cannot see how this follows from being a person. My concern is to argue against those who hold that, assuming persons do have a right to life, the early fetus has one because it is already a person.

[2]I cannot exactly say what love is, but I would argue confidently nonetheless that it is more than sexual desire. For example, it includes caring about the desired person's happiness and state of mind.

Bibliography

[1]Dennett, D. C. (1978), "Conditions of Personhood," *Brainstorms*, chapter 14, Bradford Books, Montgomery, Vermont.

[2]Engelhardt, H. T., Jr. (1976), "The Ontology of Abortion," S. Gorovitz (ed.), *Moral Problems in Medicine*, Prentice-Hall, Englewood Cliffs, New Jersey, pp. 318–334.

[3]Frankfurt, H. (1971), "Freedom of the Will and the Concept of the Person," *Journal of Philosophy* **68**, 5–20.

[4]Gallop, G. (1970), "Chimpanzees: Self-Recognition," *Science* **167**, 86–87.

[5]Hardy, J. D. (1953), "Thresholds of Pain and Reflex Contractions as Related to Noxious Stimulation," *Journal of Applied Physiology* **5**, 725–737.

[6]Hare, R. M. (1976), "Survival of the Weakest," S. Gorovitz (ed.), *Moral Problems in Medicine,* Prentice-Hall, Englewood Cliffs, New Jersey, pp. 364–369.

[7]Hick, J. (1976), *Death and the Eternal Life,* Chapter 14, Collins, London.

[8]Ingvar, D. H., *et al.* (1978), "Survival After Severe Cerebral Anoxia with Destruction of the Cerebral Cortex: the Apallic Syndrome," J. Korein (ed.), *Brain Death: Interrelated Medical and Social Issues, Annals of the New York Academy of Sciences* 315, 184–214.

[9]Korein, J. (ed.) (1978) *Brain Death: Interrelated Medical and Social Issues, Annals of the New York Academy of Sciences* 315, 142 and 366.

[10]Law Reform Commission of Canada (1979), "Criteria for the Determination of Death," *Working Paper* 23, 58–59.

[11]Parfit, D. (1976), "Rights, Interests, and Possible People," S. Gorovitz (ed.), *Moral Problems in Medicine,* Prentice-Hall, Englewood Cliffs, New J ersey, pp. 369–375.

[12]Penfield, W. (1975), *The Mystery of the Mind,* Princeton University Press, Princeton, New Jersey.

[13]Perry, J. (1978), *A Dialogue on Personal Identity and Immortality,* Hackett Publishing, Indianapolis.

[14]Puccetti, R. (1968), *Persons: A Study of Possible Moral Agents in the Universe,* chapter 1, Macmillan, London,.

[15]Tooley, M. (1976), "Abortion and Infanticide," S. Gorovitz (ed.), *Moral Problems in Medicine,* Prentice-Hall, Englewood Cliffs, New Jersey, pp. 297–317.

Human Life

vs

Human Personhood

Robert Brungs

Many are the legacies of *Roe vs Wade*, not the least of which, of course, is the death of more than ten million unborn children. Apart from the hideous slaughter there well may be one overwhelming effect of this Supreme Court decision that is not adverted to sufficiently, especially in the light of the future which already is building.

We have already consecrated in our political and social system — thanks to the seven justices who produced *Roe vs Wade*—the notion that not every human being is a person, that somehow individual human life is not the same as being a human person. The spate of movies, books, and television programs about the Holocaust rarely point out the key role of this very distinction in the deaths of millions upon millions of Jews and Slavs. We need not go back to the slavery issue to find this distinction at work. It has happened in the lifetime of many of us now living. Perhaps one of the great moral tragedies of our time is that many of those who want (rightly) to keep the horrors of the Holocaust before our minds are the same people who trivialize its meaning by encouraging the social acceptance of a distinction between human beings and human person. They seem not to hear themselves.

The rationale that permits legal abortion is that there is human life which is not protected by the State. The unborn child is not legally a person and, therefore, can be treated arbitrarily. It has no rights at all. It is the State or the society that decides who is and who is not a person under the protection of the law. In the name of reproductive *freedom* we have embarked on an essentially *totalitarian* estimate of the human being. Such a distinction has never served freedom well in the past.

Along with the distinction between individual human beings and human persons, another strange notion has been creeping into our collective psyche. How often do we hear the phrase, the right to a quality of life? In the abortion question, and especially in terms of some recent judicial decisions on wrongful birth, we get the strange notion that the right to a quality of life is somehow a more fundamental right than the right to life itself. That the quality of life depends absolutely on the fact of life seems to have escaped notice. Of course, one interpretation of this whole question of the quality of life versus life is that my right to a certain quality of life supersedes someone else's right to life. This is little more than a slightly sophisticated statement of the law of the jungle. Whether we call it the bomb shelter ethic, the life boat ethic, the tough love ethic, or any other kind of ethic, it is no more and no less an ethic of profound selfishness.

This is merely reiterating what has been said better elsewhere. But let's look at this in the context of the beginning of a whole new era in human history and in human living.

Science and Technology

Almost unnoticed by most of us—and especially by those institutions to which we have usually looked for leadership—the world has radically changed. In the past and into the present, our science, our technology, and our industry basically looked toward changes in that environment external to ourselves for our betterment. It was in this spirit that our ancestors learned to domesticate plants and animals, to irrigate lands for plant production, and to store grains and meat. It was still in that same spirit that later generations learned to harness steam and electricity and this generation is attempting to tame nuclear and thermonuclear forces. Human beings, through their understanding of the forces of nature, changed the environment in which they lived into a world which better promoted human qualities and values. That it was not always done with adequate concern for results goes without saying. But, by and large, the efforts of human beings to better their lives have been successful. The only ones who want to go back to the "good old days" are those who have forgotten what they really were like. Few of us in the highly technical societies around the world would have the skills requisite for life in the "good old days."

While the thrust of much of our science, technology, and industry has remained the same as in the past, a significantly new direction has been embarked upon. With the enormous growth (and change) in the biological sciences since about 1950, we have begun an entirely new technological adventure. We are beginning an era in which science, technology, and industry will (already do, really) look toward changes in that environment internal to ourselves, not for our betterment, but rather that we might be "better." It will be directed not to our betterment, but to our bettering! In other words, we will be among the products of our technologies and industry. The day may well come when human beings will be the most important of our artifacts. This is the challenge of this new and even novel human effort.

Thus, we are engaging in the beginning of a scientific, technological, industrial movement, much of which will impact directly and immediately on human beings. It should be pointed out that medicine, as we have known it, has been used directly and immediately on human beings, even deeply invasively. So we can legitimately ask why the new biomedical techniques should worry us at all. We have had a lot of practice over the years at the direct and immediate use of technologies on the human body and psyche. Let us, however, at once note that the new capabilities can be quite different.

In what we might call classical medicine, the techniques (whether surgical, pharmaceutical, dietary, or whatever) were used on an *ad hoc* basis to restore an individual patient to a generally recognized norm of health and/or to alleviate pain. The treatment was individualized and looked to overcoming some pathological state—guided always by the precept that the patient be no worse off after treatment than before (*primum non nocere*). Undoubtedly many of the new techniques we shall develop will pursue this same goal on a more sophisticated base. For example, we may some day be able to repair defective genes in utero. This will, when it happens, be rightly considered a major medical breakthrough. It will be one of the very significant milestones in human ingenuity, and we can all applaud the men and women who brought it about.

There will be, however, at probably the same time, the potential for "enhancing" our human genetic inheritance. When we begin to pursue this avenue of human genius, we will have to face a subtle (or perhaps not so subtle) change in medical purpose and practice. What will occur in society when the purpose of biomedical intervention into the human physical composite is no longer solely a restoration of organic health, i.e., the removal of a pathological condition? Another purpose will be surfacing, a purpose that is not at all new in the intellectual annals of our race. We shall be talking about interventions, the results of which will be transferred to future generations. That will indeed be the rub. We begin, then, to talk about eugenics, a recurring human dream (nightmare?). What will be new will be the capacity to accomplish it on a reasonably successful basis.

When we begin to "enhance" our genetic inheritance, we shall almost certainly be adding a new adjective to medical intervention. We shall not be talking only about direct and immediate interventions into human beings. If eugenics becomes a social goal, these interventions at some point will have to be *systematic* as well. Systematic carries a double implication: it must be methodical and methodological.

To achieve any eugenic effects, it will not be enough to achieve an 80 or 90 percent success. Let us take an example of negative eugenics. It has been proposed (and the proposal has met great approval in some segments of our society) that we might eliminate a genetic disease, like cystic fibrosis, by a more sophisticated amniocentetic technique and selective abortion (*Roe vs Wade*, again). If we can refine amniocentesis to reveal carriers of a defective gene (whether or not they have the genetic disease), then all the carriers of the gene would be aborted. Over a couple of generations, the disease resulting from that genetic defect would be eliminated. Of course, we are not eliminating disease, but diseased people. Anyway, it will be unproductive to kill only 80 or 90 percent of the carriers. Besides the fact that the logistics for accomplishing this are nearly impossible to achieve (every pregnancy in the world would have to be monitored amniocentetically), it is quite interesting that it is proposed. Still, the methodical attempt will be present.

We have seen this methodical aspect recently on another level. Just a few years ago the World Health Organization (WHO) was engaged in a project to eliminate smallpox from the litany of human woes. At one point they found only a handful of cases in the world, in very rural Ethiopia. Even then they were not willing to announce success until there were several months in which no cases were reported anywhere in the world. It should be noted that smallpox is an infectious disease, the elimination of which did not demand the death of those who had it. Still the WHO project shows clearly the methodical aspect of systematic "health care."

The methodical aspect of systematic technological intervention into human beings is not as important as the methodological aspect. If we are going to pursue the goal of bettering human beings, then we imply that we already know what a "good" human being is. It is here that the estimate of the human will be extremely important and where the mischief of *Roe vs Wade* will make itself felt.

We must realize that if we are going to intervene systematically into the human composite, the biosciences and biotechnologies will have to be subject to some controling view of what it means to be human. This controling view of the human is far more important and far more critical than the technologies themselves. The principal reason for any social application of these biotechnologies will, perhaps strangely enough, be more order, less randomness in the human situation. It is simply a novel form of the cry for "law and order." Any society-wide advance in systematically improving the human stock (read eugenics) will demand new criteria for judging which technologies ought to be promoted for what purposes. As we move from the concern for the well-being of individuals to concern for the well-being of society (take, for example, abortion for fetal indications and also the concern for population control), what criteria are more likely to be applied to the use of bioscientific discovery and biotechnological application? As things are going culturally and socially (*partly* thanks to the wisdom of the Supreme Court), the most likely criteria will be the basic canons of experimental science wedded to the whims and changing fads of the dominant cultural system. These canons are simplicity (efficiency, in the technological mode), predictability, and reproducibility.

Eugenics (and this is the goal of any systematic application of biosciences and biotechnologies) demands a predictably better product. Any method or technique which does not produce a predictably better product (whether animal or human) will quickly be abandoned. Without a predictable result, we might as well be content with those natural processes which have been pejoratively described as the "roulette of random reproduction." The tone of that remark seems to suggest the contempt (fear?) of anything spontaneous and uncontroled. Moreover, if the predictable results are not reproducible, eugenics is an unachievable dream; randomness will not have been reduced to order.

The canons of experimental science were developed for experimentation on inanimate objects. The adoption of the methodologies of physics into life sciences has led to the technological and industrial application of biological advance. These canons are premised on the all but total manipulability of the experimental research object. Laboratory science demands as complete a freedom as possible to transform and rearrange the structures of the experimental object. In order to obtain reproducible results, the experimenter has to be in control of the environment in which the experiment is conducted. The laboratory environment must be closed to any random, spontaneous, uncontrolled event that would affect the result. The laboratory is a closed system in which all spontaneities must be deliberately and systematically eliminated.

If the canons of experimental science do become the basis for the social application of biotechnological capability—and very many of the biotechnological proposals that are being made would have it so —the type of control needed in the laboratory would have to be imposed on the social system and, of course, on those individuals who make it up. The absence of the random, of the spontaneous, of the independent is absolutely required for reproducible results. Spontaneities such as uncontroled reproduction or any other kind of "deviant behavior"—however "deviant" might come to be defined— would not and could not be tolerated. Let us not be mistaken about this: any serious eugenics program is by nature totalitarian. Here again we see in the *Roe vs Wade* decision that "cloud as big as a man's hand." In a eugenics society, new norms of humanness will have to be created and imposed. The distinction at the heart of *Roe vs Wade*, namely, that living human beings are not legal persons until that is granted to them by an agency external to themselves, will allow normative imposition. Those who do not measure up to these new criteria can simply be designated nonpersons, since this is now the prerogative of the stronger members of society or of society itself. Ironically enough, the call to privacy as a Constitutional right, which demands the distinction between the living human beings and legal persons, will be itself one of the first victims of a eugenics program. Ironic, indeed!

We find it quite easy to assume that such laboratory conditions could never be imposed on a society as open as ours. That is probably true right now, if the attempt towards eugenics were made all at once and in all its clarity. If the attempt continues in the incremental fashion used thus far, the outcome is much more debatable. One thing is clear already: if these novel capacities are to be used to their full potential to produce predictable and reproducible results, the social system will have to be very tightly controled. There is simply no alternative! The price for social predictability and reproducibility will inevitably be the elimination of human dignity and freedom.

Each step down the biotechnological road will form the base for and be supportive of the next step down that road. That base is already being laid: the technologizing of human reproduction has already proceeded through artifical insemination, chemical contraception, abortion, in vitro fertilization, and is now looking forward to in vitro gestation. We should take a very long look at possible destinations before deciding whether to make the journey. We still have an opportunity to decide what "definition" of the human we wish to call our own. A society based on laboratory models is inevitable only if we *continue* to choose a model of the human as essentially to be malleable, objectified, transformed, and rearranged. That is a highly likely outcome, if we do not face these issues, since there is enough momentum built into technological advance to cause significant worry.

Shall we either deliberately build, or at least acquiesce in, some kind of biological collectivist society, or shall we consider the human being and his or her *inherent* dignity and freedom as paramount? There are important voices in our society, including the US Supreme Court, which are consciously or unconsciously dedicated to norms of life, dignity, and freedom external to the individual human beings. This is but one step short of giving the welfare of the group or of the race priority over the welfare of the individual. As soon as some human beings are not persons, all are at jeopardy. Certainly the new biological tools and their use can be good or evil, but their presence considerably broadens and deepens social, political, and cultural issues. We are living through the beginnings of the greatest technological revolution in the history of mankind. We stand on the threshold of the capability of deliberately directing our own future growth as a species—something absolutely novel in our history.

Let us look at one admittedly bizarre mention of "socio-genetic" engineering ideas. This is taken from an article by William Murray in the June 1975 issue of *Cosmopolitan*. While Murray's article is written in a very popular and even somewhat sensational manner, it does not really misrepresent proposals that some scientists have dumped into the public forum:

> "Here and now *Homo sapiens* is in the process of becoming *Homo biologcus,*" a noted French biologist named Dr. Jean Rostand wrote a few years ago, "a strange biped that will combine the properties of self-reproduction without males, like the green fly; of fertilizing his female at long distance, like the nautiloid mollusk; of changing sex, like the xiphores; of growing from cuttings, like the earthworm; of replacing his missing parts, like the newt; of developing outside his mothers's body, like the kangaroo; and of hibernating, like the hedgehog." And Dr. Robert C. Gesteland, an associate professor of biological sciences at Northwestern University in Illinois, has suggested crossing man with plants, so all we'd need for food would be water and sunlight; developing a servant class of super-smart apes; and best of all, breeding a race of humans only four inches tall, which would lessen pollution and conserve natural resources...Dr. Haldane (the late British geneticist) predicated we might breed, for one thing, a race of legless mutants with prehensile tails or feet for space travel. Other scientists would like to see women laying eggs that could be hatched or eaten; human beings with gills to facilitate underwater travel; people with two kinds of hands, one for heavy work, the other for lighter tasks...

Shades of the cantina scene in *Star Wars!* This sounds bizarre and is bizarre in anything like the foreseeable future. But the language is here and so are the attitudes. They are worth loking at. It is not difficult to see that *function* is the predominant rationale given for making such changes: Legless mutants *for* space travel; gills *for* underwater travel, etc. It is no exaggeration at all to say that the functional aspect of these changes is an appeal to an external principle to decide human value. In this proposed world, a human being's value would not flow from within; rather it would be determined solely by how one fits a predetermined social niche. These predictive proposals (fantasies?) look toward changing human beings to accomplish social tasks, rather than to adapt those tasks to fit human capability. When that world is upon us (and it is creeping up on us), we shall be respected and honored only for what we do, never for being what we are. In the context of another aspect of growing biotechnical achievement, I remember a cartoon, I believe, in an issue of *The American Scientist* some years ago. The scene is in a laboratory for cloning human beings. A technician is pounding on the door to the lab chief's office, shouting: "Come quick! Come quick! All the Einsteins are tap dancing." What will we do with tap dancing Einsteins? They certainly are not fulfilling the role for which their existence was planned.

But despite the difference in language, is the Supreme Court, in *Roe vs Wade* really saying anything else? It says, in effect, that you can be a living human being who is not a legal person, and has no protection under the law. This, equivalently, says that a human being has only that dignity that the State or a social consensus is willing to confer. Until this dignity is conferred, the individual can be treated (or disposed of) arbitrarily. This arbitrariness will be an essential attitude as we tinker with living human systems. It has, for example, been reported (and evidently admitted) that human embryos ("leftover human embryos") have been frozen for future use in the in vitro fertilization clinic at Queen Victoria Hospital in Melbourne, Australia. Why? Probably for the same type of experimentation that was proposed for funding by Pierre Soupart at Vanderbilt. Soupart sought funding (not granted before his death) for the study of laboratory-fertilized human embryos which, after 14 days of experiment would be killed (cf. *The Federal Register*, June 18, 1979, p. 35,039).

In the light of *Roe vs Wade*, it is quite difficult to understand why funding was not granted. If the embryo (or fetus) in utero is not granted the protection of law, if its life is so easily erasable, why worry about embryos in vitro? And, indeed, without some reverse in the mischievous concepts underlying *Roe vs Wade*, this worry will decrease. After all, we're going to need a very large rug under which to sweep the laboratory detritus. *Roe vs Wade* provides that rug—and does so, ironically enough, in the name of freedom. This has been and will continue to be true. Abortion, population control by the most efficient technical means, the culling of the weak, the retarded, the disabled, the old, and finally, positive eugenics will be proclaimed by the dominant culture *in the name of freedom*. The proponents of such measures are really "biological rednecks," biological "law and order" people, who are either fearful of what is not controlled or disdainful of what is not perfect. The sad thing is that they are seemingly unaware that the freedom they invoke cannot exist in the world they propose. Proponents of such "progress" are in reality gnostic utopians who view the human being as radically manipulable and to be transformed, whose dignity will derive solely from the success of the transformation. Charles Frankel in the March, 1974 issue of *Commentary* stated it very well:

> The most astonishing question of all posed by the advent of biomedicine, probably, is why adults of high intelligence and considerable education so regularly give themselves, on slight and doubtful provocation, to unbounded plans for remaking the race. The factor responsible is not biomedicine; something else can be the catalyst tomorrow. It is the larger idea which has shaped the major traumatic events of the last three hundred years of modern history. What unites the Puritan radicals, the Jacobins, the Bolsheviks, the Nazis, and the Maoists is the deliberate intention to create a "new man," to redo the human creature by design...
>
> The partisans of large scale eugenics planning, the Nazis aside, have usually been people of notable humanitarian sentiments. They seem not to hear themselves. It is that other music that they hear, the music that says there shall be nothing random in the world, nothing independent, nothing moved by its own vitality, nothing out of keeping with some idea: even our children must not be our progeny, but our creation.

Roe vs Wade, with its unfounded and very dangerous distinction between human beings and human persons will play a major role in the development of attempts at a eugenic society. In such a society—as history and common sense both show—neither freedom nor privacy can be allowed. This distinction, then, will go a long way toward destroying what it hoped to achieve. As our biotechnological capability increases, so too will the mischief of the Supreme Court's decision. It will guarantee the elimination of the biologically suspect. And we all—every one of us with no exceptions—are biologically suspect. So "raw judicial power" will finally yield to "raw biological power." We shall all be the losers.

Kinds and Persons

Peter A. French

1. In "The Corporation as a Moral Person,"[1] I argued that certain business corporations may be properly conceived of as intentional agents.

Corporations, of course, are artifacts. If they are to be treated as citizens of the moral world as they are in the legal world, then one of two things must be the case; (1) "person" names a natural kind or a concept that is necessarily, if in part, defined in terms of natural kinds, but "moral person" does not name a concept that depends on that person; or (2) "person" does not name a natural kind or a classificatory concept that is partially defined in terms of natural kinds and "moral person" either is, as in (1), independent of the concept of person or is interdependent with "person" or, alternatively personhood is a necessary precondition of moral personhood. "Person" clearly cannot both be the name of a natural kind and being a person be interdependent with or a necessary precondition of being a moral person, if corporations can be moral persons. But is "person" a natural kind term, or does it name a classificatory concept that is necessarily defined in terms of natural kinds?

2. A natural kind is a class of individuals gathered according to a sameness criteria that is rooted in a comprehensive scientific theory.[2] Putnam writes, "What really distinguishes the classes we count as natural kinds is itself a matter of (high level and very abstract) scientific investigation."[3] In other words, the essential nature of a thing, by virtue of which it is included in the extension of a natural kind term, that which it shares with other members of the kind, is, at base, a matter of physical theory construction.

the extension of terms. The reference of natural kind terms is fixed by ostending exemplars rather than by stereotypical characterization. Locke told a marvelous story about Adam confronted by one of his children bearing a glittering substance. Adam names the substance *Zahab* (gold). Locke says, with respect to the sameness criteria for gold,

> he (Adam) has a standard made by nature....He takes care that his idea be conformable to this Archetype and intends the name should stand for an idea so conformable. This piece of matter, thus denominated *Zahab* by Adam, being quite different from any he had seen before, nobody, I think, will deny to be a distinct species, and to have its peculiar essence; and that the name *Zahab* is the mark of the species and a name belonging to all things partaking in that essence. [4]

When Adam names the glittering substance *Zahab,* we may say that he intends to pick out *that* stuff, whatever it is and other stuff that is the same as it. On the Putnam-Kripke theory, Adam performed a kind of baptism. The exemplar stands in for the kind that includes it. The most comprehensive true scientific theory will reveal the essential properties of the kind. In the *Zahab* case that theory would expose the atomic structure of that glittering stuff.

A number of objections have been raised against the Putnam-Kripke program of assimilation of the treatment of natural kind terms to that of proper names, to treating them as rigid designators. (Some of the objections, especially those recently developed by Moravcsik[5] and by Dupre,[6] raise severe difficulties for Putnam's analysis.) For present purposes, however, such criticisms may be overlooked.

Our philosophical tradition certainly has *not* identified "person" as a natural kind term. Against that tradition, what would motivate the treatment of a "person" as a natural kind term? Wiggins[7] has noted that if "person" were a natural kind term, the puzzling fission and fusion examples concocted to throw doubt on various personal identity criteria could be conveniently handled by appeal to natural laws and scientific theory. Possible worlds, in which the laws of nature that are actually involved in fixing the referent of "person" do not hold, are worlds without persons. Wiggins writes:

> If person is a natural kind, then when we consider the problem of the identity of persons through change, the whole logic of the situation

must exempt us from taking into account any but the class of situations which conform to the actual laws of the actual world. For these serve, and nothing but these can serve to define the class of persons. And this seems to excuse us from allowing the spontaneous occurrences of delta formations in the consciousness of persons.[8]

If "person" were the name of a natural kind and we could show that delta formations in the stream of consciousness do violate the laws of nature that actually hold with respect to persons, these "puzzle cases" would disappear. To do that would be to show that a world in which delta formations occur is one in which the laws governing what it is to be a person in that world are radically different from those in our world. The corollary is Putnam's Twin Earth example of water. Despite the stereotypical similarities of Earth water with what is called "water" on Twin Earth, a distinct difference in molecular construction exists between the two. When it is discovered that XYZ rather than H_2O is the chemical formula of Twin Earth water, it is, *eo ipso*, learned that Twin Earth water is not water.[9]

The program, though it may appear to be promising with respect to settling long standing personal identity problems, is, however, as Wiggins will allow, fatally flawed. "Person" simply is not a natural kind term in the way "water," "tiger," and "human being" are. If "person" were a natural kind term, given the Putnam-Kripke theory, then its referent, like the referent of *Zahab,* could be fixed by ostension to an exemplar. Suppose we are all willing to grant that Carl Sagan and Ronald Reagan are persons, the class determined by ostension to Sagan or Reagan, given existing comprehensive scientific theory, however, they would not be the class of persons, but the class of human beings or *Homo sapiens.* "Person" and "Homo sapien" may usually be used to pick out the same referent, say Ronald Reagan, but "person" is neither a translation, nor a variant of a "Homo sapien." It will not do to say that a person is, whatever that, pointing to Sagan, is and whatever that is will be known after empirical investigation has revealed its microstructure. "Person is not manifestly equivalent to the real essence of homo sapiens."[10]

It may well be the case that providing the explanatory scientific model for the natural kind picked out by the "human being" is much more complicated than it is for water, but within the framework of a palpable zoological taxonomical theory, it should be possible to provide what is wanted. Putnam, in his Locke lectures balks at this. He tells us that: "We are not, realistically, going to get a detailed explanatory model for the natural kind 'human being'."[11] His reason for thinking that, as it happens, rests on his mistaken equation of "being a person" with "being a human being." He argues that the partial opacity of each of us to each other is a constitutive fact of what it is to be a person.

> Would it be possible to *love* someone, if we could actually carry out *calculations* of the form: "If I say X, the possibility is 15 percent she will react in manner Y?"....Would it be possible even to think of *oneself* as a person?[12]

Putnam, I think, is quite right about the partial opacity of each of us to each other, though his example surely does not show what transparency would be like. Is opacity, however, a constitutive fact about persons? If we treat what it is to be a person as solely a matter of convention, we could make opacity a part of the functional descriptions we agree to use to pick out members of the class. That, of course, is surely not what Putnam has in mind. At best we would have formulated a stereotypical feature without providing anything like the "scientifically palpable real essences" that make Putnam's natural kind theory persuasive, e.g., water is H_2O.[13] I think we do have a rather straightforward way of accounting for Putnam's partial opacity between persons, which will become clear presently. The point of interest here is that Putnam's shift from human beings to persons and his sighting of a fact about persons to establish the claim that the required scientific explanatory model of human being is not to be forthcoming, utterly fails his own enterprise. We should echo with the tradition, "person" is not, though "human being" is, the name of a natural kind.

Had Carl Sagan been presented to Locke's Adam in the way the hunk of glittering substance was, Adam might have used a term intending it stand for this specimen before him (and any of its kind) whatever essential kind properties it turned out to have. He may or may not have been later delighted to learn that he was also in the extension of that term that he and Carl Sagan were members of the same natural kind. Alternatively, Adam could have done something quite different when confronted with Carl Sagan. He might have invented a term (e.g., "person") intending it to name the aggregation of a certain collection of properties or characteristics, those he had associated to form a sortal concept and believing those properties to be present in the specimen, he calls Carl Sagan a person, that is, he includes Sagan in the extension of "person." On this version, Adam decides that Carl Sagan is one of that kind he has called "person," because Sagan has or evidences certain properties. (Adam, of course, could be wrong about whether Sagan actually has the required properties, but he cannot be wrong about what those properties are. This sortal concept is, after all, his creation.)

In the earlier case, Adam believes that Carl Sagan is of a kind or species, even if Adam has no idea or only a very bad idea of what the essential kind properties are. Suppose he names his exemplar of that kind, "human being." After intensive examination is conducted, it may turn out that Carl Sagan has, despite outward appearances, a radically different physiology from every other individual who has been regarded as a member of the human being kind that Adam had named by using Carl Sagan as his ostending exemplar. Suppose it is learned that Sagan landed in Adam's vicinity in a spaceship that had been launched from a destructing planet in a distant galaxy. Technically speaking, "human being" would have been annexed by Adam to Carl Sagan's kind and the rest of us would not be entitled to it, by reason of not being in its extension. Practically, of course, if these circumstances had occurred, we would initiate a new random exemplar test, declare a previous misuse that would exclude Carl Sagan from the class of human beings and operate within the parameters of accepted scientific theory. None of this, however, need affect Carl Sagan's status under the other kind classification. He still, we assume, behaves and evidences the characteristics that led Adam to earlier identify him in the extension of "person." This man from deepest space may not be a human being, but he is a person. The purely operational definition account of the membership criteria for the extension of "person" (here attributed to Adam), may offend the general conviction that whether or not something is a person ought to be more than a matter of legislation, social practice, or convenience.

3. There is a long tradition of dispute over whether intelligent extraterrestrial beings, computers and even some nonhuman animals

(remember Locke's rational parrot) belong in the extension of "person." There also has been reasonable dispute over whether certain human beings are persons. Wiggins certainly realizes that "person," despite important advantages if it did, does not name a natural kind. Wiggins' modification is that something is a person if it is an animal of a natural kind whose members perceive, feel, imagine, desire, etc. Person is "a nonbiological qualification of animal."[14] It is an animal classification that cuts across the grain of the usual zoological taxonomy and so efficiently excludes artifacts such as computers and corporations.

The motivation for insisting that to be a person is to be an animal is rather weak. It would appear to be the strongest claim Wiggins feels he can make that is consistent with scientific evidence and Putnam's natural kind theory and would have the effect of blocking delta developments in streams of consciousness. But Wiggins supports the inclusion of a natural kind element on little more than an appeal to what he takes to be widely held convictions or intuitions regarding the way robot cases should be handled. Most of us probably would be reluctant to issue civil rights to robots and we would probably say that we do not want to do so, because such artifacts ought not to be given consideration owed to living or natural things. Wiggins writes, "To have feelings, or purposes, or concerns a thing must, I think we still think, be (at least) *an animal.*"[15] Wiggins qualifies this claim by saying that he does not believe that an animal could not be artificially synthesized. And by that he means that we might build something that has feelings, purposes, and concerns, "programmed in." Such a creation, however, Wiggins insists "would not be an automation."[16] Visions of Hal of *2001: A Space Odyssey* spring to mind. Wiggins, of course, could say that physical laws would be violated by the actual creation of a Hal, or he might agree that Hal is capable of feelings, purposes, and concerns, and deny that Hal is an automation. But, whatever he does with such a case, he cannot deny that Hal has no place in a zoological taxonomy. Hal simply is not, nor would it make sense to say that he could become, a member of a natural kind.

The cold fact is that it is probably not that something is natural or living (in a biological sense) that stimulates our convictions about rights or considerations. The possession of certain kinds of capacities is crucial. Undoubtedly, we have a species bias that inclines us to root for the human being against Hal, but that ought not to be confused with our reflective consideration of the sort of treatment owed to and expected of feeling, purposive, concerned, desiring, self-motivated, project-making things that can "conceive of themselves as having a past accessible in experience-memory and a future accessible in intention."[17] The class of persons seems to be formed around just such a series of functional descriptions. On reflection, we may root for the human being against Hal just because Hal is guilty of intentionally killing a number of persons and, because he has manifested the intention to continue his murderous behavior against the remaining astronaut.

Wiggins concludes his argument for an animal element, in the concept of person, with the curious remark: "By *person* we mean a sort of animal, and for purposes of morality that (I maintain) is the best thing for us to mean...It is on pain of madness that we shall try to see ourselves as both *Homo sapiens* and something with a different principle of individuation."[18] The argument here is remarkably underdeveloped. Of course it would be madness if the two kinds in which we see ourselves were incompatible biological taxa. But that is exactly what the Lockean distinction between "same man" and "same person" is not. Furthermore, it is not unintelligible for me to imagine there may come a day when, though I will still be a human being, I will not be a person. It may now be my desire that when and if that day arrives, my life will be ended by the intervention of some person. Alternatively, I now may have altogether different wishes regarding my biological existence should I cease to be a person. I may now desire that extraordinary means be used to maintain the physiological functions of my body, even if I have forever lost the capabilities to feel, to have purposes and concerns, to remember, etc.

In Wiggins' own account, the natural kind element seems, for the most part, to be idle. In order for it to really have a significant role, it should function as the "glue" that holds together the collection of various capabilities, etc., that are listed in the functional descriptions that gather the kind. Wiggins expressly rejects such a view when he suggests the possibility that species or natural kinds "who come near enough to us"[19] ought also to be regarded as persons. "Near enough" is notoriously slippery. Wiggins, in fact, though perhaps with some chagrin, analyzes this nearness condition in terms of the functional descriptions. Hence, an animal of another species is a person if its species' typical members are thinking, intelligent beings capable of having purposes, desires, etc., and of considering "themselves as themselves the same thinking things, in different times and places."[20] It appears as if the animal element is but a nod to our convictions about the value of living organisms and a rather suspicious ad hoc way of solving fission and fusion puzzle cases by ruling them impossible. The natural laws that govern animal kinds do not permit of such deltas, etc., in the actual world.

4. The real difficulty with Wiggins' approach is that he seems unable to offer a way of avoiding conventionalism when it comes to the inclusion criteria for the extension of "person." The natural kind gambit fails to produce a Lockean real essence of person that will serve as an effective constraint on the conditions to be counted essential in the operational definitions. Hence, his position collapses into the descriptivist conventionalism he had set about to destroy. Wiggins, however, does raise the right issues with respect to any conventionalist account of person. The key question ought to be, "Are we free, conceptually speaking, to shorten or lengthen the list of capacities included on the functional definition of person?" If we are unconstrained in this, we could declare an open door policy that admits all manner of living things and many artifacts to the club as suits our social, legal, or moral tastes. But such a policy would render the concept of person about as useful as that of stuff or thing.

The kind of limitations on person kind membership that are wanted might be provided in two rather different ways. The first is to adopt the view that the extension of "person" is determined by a nominal essence of person that is an invention of our conceptual history. In effect, "person" would be treated not unlike artifact kinds. For example, a pencil is a tool for writing and a clock is a device for keeping time; actual mechanisms and construction materials may vary as long as certain functions are performed. Suppose we call any group of things collected according to a set of purely conventional, functional descriptions a custom-governed kind. There may be social or semantic laws of association involved, but nothing like Putnam's laws of nature will determine the crucial extension membership criteria for custom-governed kinds. The only check on adequacy of such criteria in the functional descriptions incorporated in the nominal essences of custom-governed kinds is the test of general usage and practice. The concern that we ought not to set the limits on what is a person solely by social convention, tradition, or convenience is, unhappily, not convincingly allayed if "person" names a custom-governed kind. Family resemblance, not unlike the sort to which Wiggins seems to have final resort, should be expected to play a dominant role in the determination of the extensions of custom-governed kind terms. Hence, kind membership may be extended despite radical structural differences among members.

The second alternative captures the notion that although natural laws cannot be used to govern membership in the extension of "person," total abandonment of any structural account is neither wise nor required. A body of empirically verifiable generalizations, such as, for example, those identified with common sense psychology, can function, with respect to the determination of person kind membership, in much the same way as natural laws govern natural kind membership in Putnam's theory.[21] Persons, on this account, are those things that behave in those ways that, by and large, are explainable by appeal to a coherent set of true empirical generalizations. The empirical generalizations that govern the extension of "person" cluster around such states, primary of which is intentionality, that can be ascribed to entities by us on the basis of observed outward manifestations or behavioral evidence and that explains the molar behavior of those entities. This account incorporates, it should be noted, a well-grounded doctrine in the law of contracts: the identification of intent with outward expression within the framework of ordinary meaning. Hence, Learned Hand writes:

> A contract is an obligation attached by the mere force of law to certain acts by parties, usually words, which ordinarily represent a known intent.[22]

And in the decision in Brant *v* California Dairies (4 Cal. 2d 128, 133-34, 48 P. 2d 13, 16, 1935), we read:

> ...it is now a settled principle of the law of contract that the outward manifestation or expression of assent is controlling.

There are, at least, three types of kinds: custom-governed kinds, natural law-governed kinds, and empirical generalizations-governed kinds. (There certainly are other useful ways to draw these distinctions.)[23] To avoid the pure conventionalism of custom-governed kinds, we should require that in the case of persons the extension of the kind term be governed by a coherent set of true empirical generalizations, as are to be found in common sense psychology. Such generalizations will be laced with probabilities and approximations, but inexactness of that sort is no stumbling block. Indeed, it is a feature that distinguishes the operative kind-entry criteria from natural laws. The test of the set of generalizations, actually accepted, of course, will lie in its explanatory power with respect to the behavior of exemplars of the kind.

Consider the following example: someone who desires an object and believes that acting in a certain way will obtain the object, *ceteris paribus,* will act in that way. The coherent set of like generalizations will, in fact, provide a behavioral elucidation of the concept of intentional action or agency. If something is a person, some of its molar behavior will be explainable by subsuming that behavior under those generalizations that express the concept of intentional agency. This then captures our common belief that to be a person is, at least, to be an intentional agent.[24]

Importantly, the account of person as an empirical generalizations-governed kind does not relegate the proper use of "person" to a matter of convention or politics. There are entry requirements that are not merely a matter of our legislation.

Recall that Putnam had offered as a constitutive fact of persons that they are partially opaque to each other. We are now in a position to explain how such a fact might be taken to be constitutive of the class of persons. The empirical generalizations that explicate the "sameness" relation for the kind are just that, generalizations. They should be expected to be laced with *ceteris paribus* and other qualification clauses, though with reference to much of the behavior of any random exemplar, they need to be projectable and universal, to support counterfactuals, and to ground explanations of past phenomena and predictions of future behavior. In other words, they must satisfy the conditions generally applied to laws of nature, though lacking assurance of application, the *ceteris paribus* nature of these generalizations, is the constitutive fact of partial opacity noted by Putnam. If I know from his expressions of desire that my neighbor wants a Triumph sports car and believes that if he takes a certain job, he will make enough money to purchase the car, I should predict that he will take that job. But if he does not, I will not regard his not doing so as sufficient grounds to remove his name from the roster of persons. I will, of course, ask him if he really wants the Triumph and if he really believed that taking the job would make it possible for him to buy the car, etc. I should want to check that I have the facts right. But, even if I do have all of the information, his choice to refuse the job may be based on psychological or other states about which I have, or even he has, no available knowledge. What we look for when we identify something as a person is that much of its behavior accords with the coherent set of generalizations we have identified. If in the case of a particular entity we find consistent departure from expectation, we should decide that the thing in question is not a person. If our generalizations were often undependable, we would have to frame a better set that would collect Carl Sagan and Ronald Reagan and Jane Fonda and me. Putnam's theory maintains that the extension of a natural kind term is exposed by means of a theoretical "sameness" relation given in terms of natural laws to a suitable exemplar. On my account the extension of a kind term like "person" is also exposed by means of a "sameness" relation to a suitable exemplar. The "sameness" relation in the case of "person," however, is explicated in terms of a coherent set of empirical generalizations rather than natural laws, and that, coincidentally, accounts for the fact that persons are partially opaque to each other and hence why we are not realistically going to succeed in the enterprise of constructing a detailed explanatory model for the kind "person."

To decide that something is a person is at least to have determined that it makes sense to redescribe some of its behavior in a way that makes true sentences that say that it acted intentionally. The empirical generalizations of common sense psychological theory provide the theoretic superstructure for such redescriptions.

"The Corporation as a Moral Person" locates grounds for redescriptions of corporate behavior by isolating the corporate intentionality aspect of corporate behavior. Business corporations, at least some of them, belong in the extension of person by virtue of standing in the appropriate "sameness" relation to Ronald Reagan, Carl Sagan, Jane Fonda, and me. What makes us all persons is not that we are made of the same stuff or have the same microstructure, but that our behavior can, by and large, be explained (and often predicted) by subsuming it under a certain specific set of empirical generalizations.

5. Corporations are to be included in the extension of a "person," but, some may argue, that provides only a *prima facie* reason to admit them to moral citizenry and that, on the basis of a number of other considerations, they should be excluded from the extension of "moral person." "Moral person," on this account, may name a kind that is gathered by some set of conventions or is gathered by an extended set of empirical generalizations beyond those that determine the extension of person. Some philosophers tell us that for a person to be morally responsible, he must possess or evidence a set of properties over and above those of persons *per se*. In order to be held responsible, for example, it is sometimes argued that a person must be capable of expressing regret or remorse (or both) and of suffering punishment.[25] Intentionality and the empirical generalizations that explicate it, it may be argued, are not rich enough to include the functional descriptions thought necessary for moral accountability. The extension of "moral person" would then be determined solely by our conceptual agreement about what properties we think an entity must have for it not to offend our sense of justice if we were to blame it or punish it for its deeds. This would, of course, render "moral person" a custom-governed kind and hence, whether or not something is a moral person would be nothing but a matter of convention. Even if it is granted that corporations are persons, because they do not or cannot evidence the identified capacities, they cannot be moral persons.

"Moral person" would, on this account, be a sortal term used for cataloging some persons. Its extension over persons is determined by an intension that relates directly to our conception of justice or accountability. Dispute over the proper list of functional descriptions no doubt will be long and contentious. With respect to those descriptions, I do not think corporations will fare as badly as may be usually supposed.

We need not, however, discuss specific capacity issues. I suggest that the capacities and capabilities usually wanted for "moral persons" can be unpacked from the set of empirical generalizations that elucidate intentional agency and that we have already identified as defining the "sameness" relation for "person." For example, consider regret. Imagine someone incapable of regret. What are we imagining? There seem to be two kinds of answers, one draws attention to an individual who acts in a hard-hearted, self-centered, anti-social, nasty manner. But what of the matter of incapacity? Someone may commonly manifest all of those traits, and it also be true of him that he possesses the capacity for regret. The second answer provides a technical analysis that supports predictions of someone's future behavior. When it is said that a person is incapable of regret in this sense, something is being said about his capacity to view himself as the someone who did certain things, to think of himself as the person who did x (where x is an untoward action) and to feel or wish that he had not done x or that x had not had certain upshots. Regret, understood along lines such as this, and when it has to do with the personal behavior of someone rather than with the feeling of disappointment or sorrow one might have about something that is out of one's control (e.g., "I regret I have but one life to give for my country."), is subsumable in the standard elucidation of intentional agency. We may formulate an empirical generalization of common sense psychological theory as: A person that has intentionally done something, x, and knows or remembers that he did x, and knows or believes that doing x by him (in the circumstances) was wrong or evil, *ceteris paribus* will manifest some distress, or sorrow, or even grief when he thinks about or remembers his having done x. Saying of someone that he is not a moral person because he cannot feel regret, then amounts to saying that he is not a person.

6. There is one response to this general program that should be mentioned. It is the gambit that claims that "moral person" is really the name of the *complete* person, the fully equipped and operational model. An incapacity of the relevant moral variety will illegitimize moral responsibility ascriptions, while not affecting the integrity of the person in matters of personal identity. It is even fashionable in some circles to refer to such morally incapacitated individuals as dysfunctional persons. Yet the notion of a dysfunctional person is as slippery as any imaginable. If it means that an individual is technically in the extension of person by satisfying the "sameness" relation to Carl Sagan but is sadly impaired in such a way as to lack some crucial ability of complete persons, the concept of person has been knocked out of joint. All manner of things might have to be treated as dysfunctional persons. Remember the story of the prince who was turned into a frog? Imagine that it was not a talking, distressed,

unhappy frog that knew it was really a prince, but just a frog, croaking away its little life on a lily pond, eating flies rapaciously, utterly oblivious to the fact that if a beautiful princess were, for God knows what reason, to kiss it, it would turn into a handsome prince. Would *that* frog qualify as a dysfunctional person, or is it not a person at all or what? If "dysfunctional person" only means that something once upon a time was a person, but now is crucially disordered so that the empirical generalizations that are true of persons are no longer true of it, then we might just as well call it a former person. If there is a reasonable use for "dysfunctional person," it could be in cases where we believe, on good grounds, that the disorder or impairment is repairable or curable. The term would then serve as a protection device against the loss of personal identity during a gap, as for example, in near complete amnesia cases and perhaps in the frog-prince case. On the other hand, if the disorder is regarded as irreversible or irreparable, there are just no good reasons to cling to "person" at all. Hard-hearted though it may sound, the popular expressions, "He's just an animal!" and (in other contexts) "He's just a vegetable!" have appropriate uses in cases of some former persons who are human beings.

7. We are, I think, best advised to hold that if something is a person, it is a moral person and reject the idea that a moral person is some special kind or variety of person, or a person with a difference, or a fully developed person, or whatever. If, as I have argued, intentional agency is the conceptual nexus elucidated by a set of empirical generalizations that define the "sameness" relation to a suitable exemplar that determines the extension of "person," and no more or less is needed to determine membership in the extension of "moral person," then, if corporations are intentional agents (as I have argued that they are),[26] they are persons *cum* moral persons. And, of course, that will also be true for Carl Sagan, Ronald Reagan, Jane Fonda, and me.

Notes

[1]Peter A. French, "The Corporation as a Moral Person," *American Philosophical Quarterly*, **16** (1979), 207–215.

[2]Hilary Putnam, "Meaning, and Reference," *The Journal of Philosophy*, **70** (1973), 699–711, reprinted in Stephen P. Schwartz, *Naming, Necessity, and Natu-*

ral Kinds (Ithaca: Cornell University Press, 1977), p. 119–132. References are to the Schwartz volume.

[3]Hilary Putnam, "Is Semantics Possible?" *Naming, Necessity, and Natural Kinds*, p.104.

[4]John Locke, *An Essay Concerning Human Understanding* (1690), Book III, Chapter VI.

[5]Julius Moravcsik, "How Do Words Get Their Meaning?" *The Journal of Philosophy*, **78** (1981), 5-24

[6]John Dupre, "Natural Kinds and Biological Taxa," *Philosophical Review* (1981), 66–90.

[7]David Wiggins, "Locke, Butler and the Stream of Consciousness: And Men as a Natural Kind" *The Identities of Persons*, Amelie Rorty, ed., (Berkeley: University of California Press, 1976), pp. 139–173.

[8]*Ibid.*, p. 158.

[9]Putnam, "Meaning and Reference," op. cit.

[10]Wiggins, op. cit., p. 160.

[11]Hilary Putnam, *Meaning and the Moral Sciences* (London: Routledge & Kegan Paul, 1978), p. 62.

[12]*Ibid.*, p. 63.

[13]*See* Putnam, "Meaning and Reference," for a full account of this example.

[14]Wiggins, op. cit., p. 161.

[15]*Ibid.*, p. 162

[16]*Ibid.*

[17]*Ibid.*, p. 161.

[18]*Ibid.*, p. 167.

[19]*Ibid .*, p. 161.

[20]*Ibid.*

[21]This approach is suggested and defended by Patricia Kitcher, "Natural Kinds and Unnatural Persons," *Philosophy*, **54** (1979), 541–547.

[22]*Hotchkiss v The National City Bank*, 200 F. 287 at 293 (SDNY, 1911).

[23]One particularly fruitful way may be to do so in terms of what Moravcsik calls aitiational frames. *See* Moravcsik, *op. cit.*

[24]*See* French, The Scope of Morality (chap. 1). Also Donald Davidson "Agency," *Agent, Action, and Reason*, ed. Binkley, Bronaugh, and Marros (Toronto: Toronto University Press, 1971), pp. 26–37.

[25]As an example *see* John Danley, "Corporate Moral Agency: The Case for Anthropological Bigotry," Bowling Green State University Conference in Applied Philosophy, 1980.

[26]*See* footnote 1.

Selected Bibliography

Adams, E. M. "Personhood and Human Rights," *Man and World* **8,** Fall 1975.

Adams, E. M. "Persons and Morality," *Philosophy and Phenomenological Research* **42,** March 1982.

Aldrich, Virgil "On What it is Like to Be a Man," *Inquiry* **16,** 1973

Annis, David "Self-Conscience and the Right to Life," *Southwest Journal of Philosophy* **6,** Summer 1975.

Arbib, Michael A. *In Search of the Person* (University of Massachusetts Press, 1985).

Armstrong, Robert "The Right to Life," *Journal of Social Philosophy* **8,** January 1977.

Bajema, Clifford E. *Abortion and the Meaning of Personhood* (Grand Rapids: Baker Book House, 1976).

Baron, Charles H. " The Concept of Person in the Law," Shaw and Doudera, 1983.

Beauchamp, Tom L. and James F. Childress, eds., *Principles of Biomedical Ethics,* (Oxford University Press, 1979).

Beauchamp, Tom L. and LeRoy Walters, eds., *Contemporary Issues in Bioethics* (Wadsworth Publishing Co., 1981).

Becker, Lawrence C. "Human Being: The Boundaries of the Concept," *Philosophy and Public Affairs* **4,** 4, Summer 1975.

Benjamin, Martin and Joy Curtis *Ethics in Nursing* (Oxford University Press, 1981).

Bertocci, Peter A. "The Essence of a Person," *Monist* **61,** January 1978.

Bertocci, Peter A. "The Person, His Personality and Environment," *Review of Metaphysics* **32,** June 1979.

Black, Max "Humaness," *The Prevalence of Humbug and Other Essays* (Cornell University Press, 1983).

Bok, Sissela "Ethical Problems of Abortion," *Hastings Center Report,* vol. 2, January 1974.

Bok, Sissela "Who Shall Count as a Human Being?" Robert L. Perkins, ed., 1974.

Bondeson, W. B. et al., eds., *Abortion and the Status of the Fetus* (D. Reidel Publishing Co., 1983).

Brody, Baruch *Abortion and the Sanctity of Human Life: A Philosophical View* (MIT Press, 1975).

Brody, Baruch "On the Humanity of the Fetus," Robert L. Perkins, ed., 1974.

Brody, Howard *Ethical Decisions in Medicine* (Little, Brown and Co., 1981).

Brungs, Robert "Human Life vs Human Personhood," *Human Life Review,* Summer 1972.

Bukala, C. R. "The Existential Structure of Person," *Personalist* **49,** Spring 1968.

Callahan, Daniel *Abortion: Law, Choice and Morality* (Macmillan, 1970).

Carrier, L. S. "Abortion and the Right to Life," *Social Theory and Practice* **3,** Fall 1975.

Carter, W. R. "Death and Body Transfiguration," *Mind* **92,** 371, 1981.

Carter, W. R. "Do Zygotes Become People?" *Mind* **93,** January 1982.

Carter, W. R. "Once and Future Persons," *American Philosophical Quarterly* **17,** 1, January 1980.

Cheng, Charles L. Y. "On Puccetti's Two-Person View of Man," *Southern Journal of Philosophy* **16,** Spring 1978.

Clack, Robert J. "The Myth of the Conscious Robot," *Personalist* **49,** Summer 1968.

Clouser, K. Danner "Biomedical Ethics: Some Reflections and Exhortations," *Monist* **60,** January 1977.

Coburn, Robert C. "Persons and Psychological Concepts," *American Philosophical Quarterly* **4,** 3, July 1967.

Daniels, Charles "Abortion and Potential," *Dialogue* **18,** June 1979.

Daniels, Norman "Moral Theory and the Plasticity of Persons," *Monist* **62,** July 1979.

Danto, Arthur C. "Persons," *Encyclopedia of Philosophy,* vol. 6, Paul Edwards, ed. (Free Press, 1967).

Davis, John W. et al., eds., *Contemporary Issues in Biomedical Ethics* (Humana Press, 1978).

Delattre, Edwin "Rights, Responsibilities, and Future Persons," *Ethics* **82,** April 1972.

Dennett, Daniel C. "Conditions of Personhood," *Identities of Persons,* A. O. Rorty, ed. (University of California Press, 1976).

Dennett, Daniel C. "Where Am I?" *Brainstorms* (MIT Press, 1981).

Deutch, Eliot *Personhood, Creativity and Freedom* (Honolulu: University Press of Hawaii, 1982).

Downie, R. S. and Elizabeth Telpher *Respect for Persons* (New York: Schocken Books, 1969).

Duska, Ronald "On Confusing Human Beings and Persons," *Proceedings of the American Catholic Philosophical Association,* vol. LIX, 1984.

Engelhardt Jr., H. Tristram "Medicine and the Concept of Person," Beauchamp and Walters, op. cit.

Engelhardt Jr., H. Tristram "Ontology of Abortion," *Ethics* **84,** April 1974.

English, Jane "Abortion and the Concept of a Person," *Canadian Journal of Philosophy* **5,** 2, October 1975.

Feinberg, Joel, ed., *The Problem of Abortion,* 2nd ed. (Wadsworth Publishing Co., 1984).

Fletch, Leonard M. "Abortion, Deformed Fetuses, and the Omega Pill," *Philosophical Studies* **36,** October 1979.

Fletcher, Joseph *Humanhood: Essays in Biomedical Ethics* (Prometheus Books, 1979).

Fletcher, Joseph "Indicators of Humanhood: A Tentative Profile of Man," *Hastings Center Report,* vol. 2, November 1972.

Frankfurt, Harry C. "Freedom of the Will and the Concept of a Person," *Journal of Philosophy* **LXVIII,** 1, January 14, 1971.

French, Peter "Kinds and Persons," *Philosophy and Phenomenological Research* **44,** 2, December 1983.

Fromer, Margot Joan *Ethical Issues in Health Care* (C. V. Mosby Co., 1981).

Glantz, Leonard "Is the Fetus a Person? A Lawyer's View," *see* Bondeson, 1983.

Glover, Jonathan *Causing Death and Saving Lives* (Penguin Books, 1977).

Gorovitz, Samuel, et al., eds., *Moral Problems in Medicine* (Prentice-Hall Publishing Co., 1976).

Gorovitz, Samuel *Doctor's Dilemmas* (Oxford University Press, 1982).

Govier, Trudy "What Should We Do About Future People?" *American Philosophical Quarterly* **16,** 2, 1979.

Green, O. H., ed., Respect for Persons, vol. XXXI of *Tulane Studies in Philosophy.*

Grobstein, Clifford "A Biological Perspective on the Origin of Human Life and Personhood," Shaw and Doudera, 1983.

Hampshire, Stuart *Thought and Action* (University of Notre Dame Press, 1982).

Haugeland, John "Heidegger on Being a Person," *Nous* **16,** March 1982.

Haworth, Lawrence "Dworkin, Rights, and Persons," *Canadian Journal of Philosophy* **9,** Spring 1979.

Jeffko, Walter G. "Action, Personhood, and Fact-Value," *Thomist* **40,** January 1976.

Joyce, Robert E. "Personhood and the Conception Event," *The New Scholasticism* **52,** Winter 1978.

Kohl, Marvin, ed., *Infanticide and the Value of Life* (Prometheus Books, 1975).

Langerak, Edward A. "Abortion: Listening to the Middle," *Hastings Center Report,* vol. 9, October 1979.

Locke, John *An Enquiry Concerning Human Understanding,* P. H. Nidditch, ed. (Oxford University Press, 1975). *See* esp. book 11, chapter XXV, "Of Identity and Diversity".

Lomasky, Loren E. "Being a Person—Does it Matter?" *Philosophical Topics* **12,** 3, 1982.

Macklin, Ruth "When Human Rights Conflict: Two Persons, One Body," *see* Shaw and Doudera, 1983.

Margolis, Joseph *Philosophy of Psychology* (Prentice-Hall Publishing Co., 1984).

May, William E. "What Makes a Human Being to be a Being of Moral Worth?" *Thomist* **40,** July 1976.

McCarthy, Donald G., ed., *Beginning of Personhood: A Symposium* (Houston: The Institute of Religion and Human Development, 1973).

McLaughlin, R. J. "Men, Animals and Personhood," *Proceedings of the American Catholic Philosophical Association,* vol. LIX, 1984.

McMullin, Ernan "Persons in the Universe," *Zygon* **15,** March 1980.

Melden, I. A. *Rights and Persons* (University of California Press, 1977).

Moraczewski, Albert S. "Human Personhood: A Study in Personalized Biology," Bondeson, 1983.

Munson, Ronald, ed., *Intervention and Reflection,* 2nd ed. (Wadsworth Publishing Co., 1983).

Nagel, Thomas "Armstrong on the Mind," *Philosophical Review* **79,** July 1970.

Nagel, Thomas "What is it Like to Be a Bat?" *Philosophical Review,* October 1974.

Newton, Lisa "Humans and Persons: A Reply to Tristram Engelhardt," *Ethics* **85,** July 1975

Noonan Jr., John T. *The Morality of Abortion* (Harvard University Press, 1970).

Perkins, Robert L., ed., *Abortion: Pro and Con* (Schenkman Publishing Co., 1974).

Perkoff, Gerald T. "Toward a Normative Definition of Personhood," *see* Bondeson, 1983.

Pinchard, Terry "Models of the Person," *Canadian Journal of Philosophy* **10,** December 1980.

Puccetti, Roland "The Life of a Person," *see* Bondeson, 1983.

Putnam, Hilary "Robots: Machines or Artificially Created Life," *see Modern Materialism,* J. O'Connor, ed. (New York, 1969).

Ramsey, Paul *The Patient as Person* (Yale University Press, 1970).

Ray, A. Chadwick "Humanity, Personhood and Abortion," *International Philosophical Quarterly* **25,** 3, September 1985.

Regan, Tom "Feinberg on What Sorts of Beings Can Have Rights," *Southern Journal of Philosophy* **14,** Winter 1976.

Regan, Tom, ed., *Matters of Life and Death* (Random House, 1980).

Reichmann, James B. *Philosophy of the Human Person* (Loyola University Press, 1986).

Reilly, Richard "Will and the Concept of a Person," *Proceedings of the American Catholic Philosophical Association,* vol. 53, 1979.

Reinelt, Herbert R. "The Nature and Knowledge of Persons," *Pacific Philosophical Forum* **6,** May, 1968.

Roca, Octavio "Atoms and Persons," *Philosophical Forum* **12,** Fall 1980.

Rorty, Richard *Philosophy and the Mirror of Nature* (Princeton University Press, 1979). *See* esp. chapter 3, "Persons Without Minds."

Ryan Jr., George M. "Medical Implications of Bestowing Personhood on the Unborn," Shaw and Doudera, 1983.

Sendaydiego, Henry B. "I am Who/What I am and More," *Journal of the West Virginia Philosophical Society* **17,** Fall 1979

Shaffer, Jerome A. *Philosophy of Mind* (Prentice-Hall Publishing Co., 1968).

Shaw, M. W. and Doudera, A. E., eds., *Defining Human Life: Medical, Legal, and Ethical Implications* (Ann Arbor: Health Administration Press, 1983).

Sherwin, Susan "The Concept of a Person in the Context of Abortion," *Bioethics Quarterly* **3,** Spring 1981.

Singer, Peter *Practical Ethics* (Cambridge University Press, 1979).

Slote, Michael *Metaphysics and Essence* (New York University Press, 1975).

Smith, Wilfred C. "Thinking About Persons," *Humanitas* **15,** May 1979.

Strawson, P. F. *Individuals* (London: Methuen and Co., 1971).

Taylor, Richard *Metaphysics,* 3rd ed. (Prentice-Hall Publishing Co., 1983).

Teichman, Jenny "The Definition of Person," *Philosophy* **60,** 232, April 1985.

Thomas, Claire "Potential for Personhood," *Bioethics Quarterly* **2,** Fall 1980.

Thomson, Judith Jarvis "A Defense of Abortion," *Philosophy and Public Affairs* **1,** 1, Fall 1971.

Tooley, Michael "Abortion and Infanticide," *Philosophy and Public Affairs* **2**, 1, Fall 1972.

Tooley, Michael *Abortion and Infanticide* (Oxford University Press, 1983).

Tooley, Michael "A Defense of Abortion and Infanticide", *see* Feinberg, 1984.

Van De Vate Jr., Dwight "Violence and Persons", *Philosophical Forum*, **7**, March 1969.

van Melsen, A. G. M. "Person", vol. 3 of *Encyclopedia of Bioethics*, W. T. Reich, ed. (Free Press, 1978).

Van Straaten, Zak, ed. *Philosophical Subjects* (Oxford University Press, 1980).

Van De Vate Jr., Dwight "Violence and Persons," *Philosophical Forum* **7**, March 1969.

van Melsen, A. G. M. "Person," vol. 3 of *Encyclopedia of Bioethics*, W. T. Reich, ed. (Free Press, 1978).

Van Straaten, Zak, ed., *Philosophical Subjects* (Oxford University Press, 1980).

Veatch, Robert M. *A Theory of Medical Ethics* (Basic Books, 1981).

Wade, Francis C. "Potentiality in the Abortion Discussion," *Review of Metaphysics* **29**, December 1975.

Warren, Mary Anne "Do Potential People Have Moral Rights?" *Canadian Journal of Philosophy* **7**, June 1977.

Warren, Mary Anne "On the Moral and Legal Status of Abortion," *Monist* **57**, 1973.

Weiss, Roslyn "The Perils of Personhood," *Ethics* **89**, October 1978.

Werhane, Patricia H. *Persons, Rights and Corporations* (Prentice-Hall Publishing Co., 1985).

Werner, Richard "Abortion: The Moral Status of the Unborn," *Social Theory and Practice* **3**, Fall 1974

Wertheimer, Roger "Understanding the Abortion Argument," *Philosophy and Public Affairs* **1**, 1, Fall 1971.

Wikler, Daniel I. "Concepts of Personhood: A Philosophical Perspective," *see* Shaw and Doudera, 1983.

Wikler, Daniel I. "Ought We To Try to Save Aborted Fetuses?" *Ethics* **90**, October 1979.

Wilson, David C. "Functionalism and Moral Personhood: One View Considered," *Philosophy and Phenomenological Research* **44**, 4, June 1984.

Young, Frederic C. "On Dennett's Conditions of Personhood," *Auslegung* **6**, June 1979.

Index